Food Safety Act 1990

1990 CHAPTER 16

An Act to make new provision in place of the Food Act 1984 (except Parts III and V), the Food and Drugs (Scotland) Act 1956 and certain other enactments relating to food; to amend Parts III and V of the said Act of 1984 and Part I of the Food and Environment Protection Act 1985; and for connected purposes.   [29th June 1990]

B E IT ENACTED by the Queen's most Excellent Majesty, by and with the advice and consent of the Lords Spiritual and Temporal, and Commons, in this present Parliament assembled, and by the authority of the same, as follows:—

PART I

PRELIMINARY

1.—(1) In this Act "food" includes—

    (a) drink;

    (b) articles and substances of no nutritional value which are used for human consumption;

    (c) chewing gum and other products of a like nature and use; and

    (d) articles and substances used as ingredients in the preparation of food or anything falling within this subsection.

(2) In this Act "food" does not include—

    (a) live animals or birds, or live fish which are not used for human consumption while they are alive;

    (b) fodder or feeding stuffs for animals, birds or fish;

    (c) controlled drugs within the meaning of the Misuse of Drugs Act 1971; or

    (d) subject to such exceptions as may be specified in an order made by the Ministers—

Meaning of "food" and other basic expressions.

1971 c. 38.

(i) medicinal products within the meaning of the Medicines Act 1968 in respect of which product licences within the meaning of that Act are for the time being in force; or

(ii) other articles or substances in respect of which such licences are for the time being in force in pursuance of orders under section 104 or 105 of that Act (application of Act to other articles and substances).

(3) In this Act, unless the context otherwise requires—

"business" includes the undertaking of a canteen, club, school, hospital or institution, whether carried on for profit or not, and any undertaking or activity carried on by a public or local authority;

"commercial operation", in relation to any food or contact material, means any of the following, namely—

(a) selling, possessing for sale and offering, exposing or advertising for sale;

(b) consigning, delivering or serving by way of sale;

(c) preparing for sale or presenting, labelling or wrapping for the purpose of sale;

(d) storing or transporting for the purpose of sale;

(e) importing and exporting;

and, in relation to any food source, means deriving food from it for the purpose of sale or for purposes connected with sale;

"contact material" means any article or substance which is intended to come into contact with food;

"food business" means any business in the course of which commercial operations with respect to food or food sources are carried out;

"food premises" means any premises used for the purposes of a food business;

"food source" means any growing crop or live animal, bird or fish from which food is intended to be derived (whether by harvesting, slaughtering, milking, collecting eggs or otherwise);

"premises" includes any place, any vehicle, stall or moveable structure and, for such purposes as may be specified in an order made by the Ministers, any ship or aircraft of a description so specified.

(4) The reference in subsection (3) above to preparing for sale shall be construed, in relation to any contact material, as a reference to manufacturing or producing for the purpose of sale.

Extended meaning of "sale" etc.

2.—(1) For the purposes of this Act—

(a) the supply of food, otherwise than on sale, in the course of a business; and

(b) any other thing which is done with respect to food and is specified in an order made by the Ministers,

shall be deemed to be a sale of the food, and references to purchasers and purchasing shall be construed accordingly.

(2) This Act shall apply—

   (a) in relation to any food which is offered as a prize or reward or given away in connection with any entertainment to which the public are admitted, whether on payment of money or not, as if the food were, or had been, exposed for sale by each person concerned in the organisation of the entertainment;

   (b) in relation to any food which, for the purpose of advertisement or in furtherance of any trade or business, is offered as a prize or reward or given away, as if the food were, or had been, exposed for sale by the person offering or giving away the food; and

   (c) in relation to any food which is exposed or deposited in any premises for the purpose of being so offered or given away as mentioned in paragraph (a) or (b) above, as if the food were, or had been, exposed for sale by the occupier of the premises;

and in this subsection "entertainment" includes any social gathering, amusement, exhibition, performance, game, sport or trial of skill.

3.—(1) The following provisions shall apply for the purposes of this Act.

(2) Any food commonly used for human consumption shall, if sold or offered, exposed or kept for sale, be presumed, until the contrary is proved, to have been sold or, as the case may be, to have been or to be intended for sale for human consumption.

Presumptions that food intended for human consumption.

(3) The following, namely—

   (a) any food commonly used for human consumption which is found on premises used for the preparation, storage, or sale of that food; and

   (b) any article or substance commonly used in the manufacture of food for human consumption which is found on premises used for the preparation, storage or sale of that food,

shall be presumed, until the contrary is proved, to be intended for sale, or for manufacturing food for sale, for human consumption.

(4) Any article or substance capable of being used in the composition or preparation of any food commonly used for human consumption which is found on premises on which that food is prepared shall, until the contrary is proved, be presumed to be intended for such use.

4.—(1) In this Act—

   "the Minister" means, subject to subsection (2) below—

      (a) in relation to England and Wales, the Minister of Agriculture, Fisheries and Food or the Secretary of State;

      (b) in relation to Scotland, the Secretary of State;

   "the Ministers" means—

      (a) in relation to England and Wales, the following Ministers acting jointly, namely, the Minister of Agriculture, Fisheries and Food and the Secretaries of State respectively concerned with health in England and food and health in Wales;

      (b) in relation to Scotland, the Secretary of State.

Ministers having functions under Act.

(2) In this Act, in its application to emergency control orders, "the Minister" means the Minister of Agriculture, Fisheries and Food or the Secretary of State.

Food authorities and authorised officers.

5.—(1) Subject to subsections (3) and (4) below, the food authorities in England and Wales are—

    (a) as respects each London borough, district or non-metropolitan county, the council of that borough, district or county;

    (b) as respects the City of London (including the Temples), the Common Council;

    (c) as respects the Inner Temple or the Middle Temple, the appropriate Treasurer.

(2) Subject to subsection (3)(a) below, the food authorities in Scotland are the islands or district councils.

(3) Where any functions under this Act are assigned—

1984 c. 22.

1897 c. 38.

    (a) by an order under section 2 or 7 of the Public Health (Control of Disease) Act 1984, to a port health authority or, by an order under section 172 of the Public Health (Scotland) Act 1897, to a port local authority;

1936 c. 49.

    (b) by an order under section 6 of the Public Health Act 1936, to a joint board for a united district; or

1985 c. 51.

    (c) by an order under paragraph 15(6) of Schedule 8 to the Local Government Act 1985, to a single authority for a metropolitan county,

any reference in this Act to a food authority shall be construed, so far as relating to those functions, as a reference to the authority to whom they are so assigned.

(4) The Ministers may by order provide, either generally or in relation to cases of a particular description, that any functions under this Act which are exercisable concurrently—

    (a) as respects a non-metropolitan district, by the council of that district and the council of the non-metropolitan county;

    (b) as respects the Inner Temple or the Middle Temple, by the appropriate Treasurer and the Common Council,

shall be exercisable solely by such one of those authorities as may be specified in the order.

(5) In this section—

    "the appropriate Treasurer" means the Sub-Treasurer in relation to the Inner Temple and the Under Treasurer in relation to the Middle Temple;

    "the Common Council" means the Common Council of the City of London;

    "port local authority" includes a joint port local authority.

(6) In this Act "authorised officer", in relation to a food authority, means any person (whether or not an officer of the authority) who is authorised by them in writing, either generally or specially, to act in matters arising under this Act; but if regulations made by the Ministers so provide, no person shall be so authorised unless he has such qualifications as may be prescribed by the regulations.

6.—(1) In this Act "the enforcement authority", in relation to any provisions of this Act or any regulations or orders made under it, means the authority by whom they are to be enforced and executed.

PART I
Enforcement of
Act.

(2) Every food authority shall enforce and execute within their area the provisions of this Act with respect to which the duty is not imposed expressly or by necessary implication on some other authority.

(3) The Ministers may direct, in relation to cases of a particular description or a particular case, that any duty imposed on food authorities by subsection (2) above shall be discharged by the Ministers or the Minister and not by those authorities.

(4) Regulations or orders under this Act shall specify which of the following authorities are to enforce and execute them, either generally or in relation to cases of a particular description or a particular area, namely—

(a) the Ministers, the Minister, food authorities and such other authorities as are mentioned in section 5(3) above; and

(b) in the case of regulations, the Commissioners of Customs and Excise;

and any such regulations or orders may provide for the giving of assistance and information, by any authority concerned in the administration of the regulations or orders, or of any provisions of this Act, to any other authority so concerned, for the purposes of their respective duties under them.

(5) An enforcement authority in England and Wales may institute proceedings under any provisions of this Act or any regulations or orders made under it and, in the case of the Ministers or the Minister, may take over the conduct of any such proceedings which have been instituted by some other person.

PART II

MAIN PROVISIONS

Food safety

7.—(1) Any person who renders any food injurious to health by means of any of the following operations, namely—

Rendering food
injurious to
health.

(a) adding any article or substance to the food;

(b) using any article or substance as an ingredient in the preparation of the food;

(c) abstracting any constituent from the food; and

(d) subjecting the food to any other process or treatment,

with intent that it shall be sold for human consumption, shall be guilty of an offence.

(2) In determining for the purposes of this section and section 8(2) below whether any food is injurious to health, regard shall be had—

(a) not only to the probable effect of that food on the health of a person consuming it; but

(b) also to the probable cumulative effect of food of substantially the same composition on the health of a person consuming it in ordinary quantities.

PART II

(3) In this Part "injury", in relation to health, includes any impairment, whether permanent or temporary, and "injurious to health" shall be construed accordingly.

Selling food not complying with food safety requirements.

8.—(1) Any person who—

(a) sells for human consumption, or offers, exposes or advertises for sale for such consumption, or has in his possession for the purpose of such sale or of preparation for such sale; or

(b) deposits with, or consigns to, any other person for the purpose of such sale or of preparation for such sale,

any food which fails to comply with food safety requirements shall be guilty of an offence.

(2) For the purposes of this Part food fails to comply with food safety requirements if—

(a) it has been rendered injurious to health by means of any of the operations mentioned in section 7(1) above;

(b) it is unfit for human consumption; or

(c) it is so contaminated (whether by extraneous matter or otherwise) that it would not be reasonable to expect it to be used for human consumption in that state;

and references to such requirements or to food complying with such requirements shall be construed accordingly.

(3) Where any food which fails to comply with food safety requirements is part of a batch, lot or consignment of food of the same class or description, it shall be presumed for the purposes of this section and section 9 below, until the contrary is proved, that all of the food in that batch, lot or consignment fails to comply with those requirements.

(4) For the purposes of this Part, any part of, or product derived wholly or partly from, an animal—

(a) which has been slaughtered in a knacker's yard, or of which the carcase has been brought into a knacker's yard; or

(b) in Scotland, which has been slaughtered otherwise than in a slaughterhouse,

shall be deemed to be unfit for human consumption.

(5) In subsection (4) above, in its application to Scotland, "animal" means any description of cattle, sheep, goat, swine, horse, ass or mule; and paragraph (b) of that subsection shall not apply where accident, illness or emergency affecting the animal in question required it to be slaughtered as mentioned in that paragraph.

Inspection and seizure of suspected food.

9.—(1) An authorised officer of a food authority may at all reasonable times inspect any food intended for human consumption which—

(a) has been sold or is offered or exposed for sale; or

(b) is in the possession of, or has been deposited with or consigned to, any person for the purpose of sale or of preparation for sale;

and subsections (3) to (9) below shall apply where, on such an inspection, it appears to the authorised officer that any food fails to comply with food safety requirements.

(2) The following provisions shall also apply where, otherwise than on such an inspection, it appears to an authorised officer of a food authority that any food is likely to cause food poisoning or any disease communicable to human beings.

(3) The authorised officer may either—

    (a) give notice to the person in charge of the food that, until the notice is withdrawn, the food or any specified portion of it—

        (i) is not to be used for human consumption; and

        (ii) either is not to be removed or is not to be removed except to some place specified in the notice; or

    (b) seize the food and remove it in order to have it dealt with by a justice of the peace;

and any person who knowingly contravenes the requirements of a notice under paragraph (a) above shall be guilty of an offence.

(4) Where the authorised officer exercises the powers conferred by subsection (3)(a) above, he shall, as soon as is reasonably practicable and in any event within 21 days, determine whether or not he is satisfied that the food complies with food safety requirements and—

    (a) if he is so satisfied, shall forthwith withdraw the notice;

    (b) if he is not so satisfied, shall seize the food and remove it in order to have it dealt with by a justice of the peace.

(5) Where an authorised officer exercises the powers conferred by subsection (3)(b) or (4)(b) above, he shall inform the person in charge of the food of his intention to have it dealt with by a justice of the peace and—

    (a) any person who under section 7 or 8 above might be liable to a prosecution in respect of the food shall, if he attends before the justice of the peace by whom the food falls to be dealt with, be entitled to be heard and to call witnesses; and

    (b) that justice of the peace may, but need not, be a member of the court before which any person is charged with an offence under that section in relation to that food.

(6) If it appears to a justice of the peace, on the basis of such evidence as he considers appropriate in the circumstances, that any food falling to be dealt with by him under this section fails to comply with food safety requirements, he shall condemn the food and order—

    (a) the food to be destroyed or to be so disposed of as to prevent it from being used for human consumption; and

    (b) any expenses reasonably incurred in connection with the destruction or disposal to be defrayed by the owner of the food.

(7) If a notice under subsection (3)(a) above is withdrawn, or the justice of the peace by whom any food falls to be dealt with under this section refuses to condemn it, the food authority shall compensate the owner of the food for any depreciation in its value resulting from the action taken by the authorised officer.

(8) Any disputed question as to the right to or the amount of any compensation payable under subsection (7) above shall be determined by arbitration.

(9) In the application of this section to Scotland—

    (a) any reference to a justice of the peace includes a reference to the sheriff and to a magistrate;

    (b) paragraph (b) of subsection (5) above shall not apply;

    (c) any order made under subsection (6) above shall be sufficient evidence in any proceedings under this Act of the failure of the food in question to comply with food safety requirements; and

    (d) the reference in subsection (8) above to determination by arbitration shall be construed as a reference to determination by a single arbiter appointed, failing agreement between the parties, by the sheriff.

Improvement notices.

10.—(1) If an authorised officer of an enforcement authority has reasonable grounds for believing that the proprietor of a food business is failing to comply with any regulations to which this section applies, he may, by a notice served on that proprietor (in this Act referred to as an "improvement notice")—

    (a) state the officer's grounds for believing that the proprietor is failing to comply with the regulations;

    (b) specify the matters which constitute the proprietor's failure so to comply;

    (c) specify the measures which, in the officer's opinion, the proprietor must take in order to secure compliance; and

    (d) require the proprietor to take those measures, or measures which are at least equivalent to them, within such period (not being less than 14 days) as may be specified in the notice.

(2) Any person who fails to comply with an improvement notice shall be guilty of an offence.

(3) This section and section 11 below apply to any regulations under this Part which make provision—

    (a) for requiring, prohibiting or regulating the use of any process or treatment in the preparation of food; or

    (b) for securing the observance of hygienic conditions and practices in connection with the carrying out of commercial operations with respect to food or food sources.

Prohibition orders.

11.—(1) If—

    (a) the proprietor of a food business is convicted of an offence under any regulations to which this section applies; and

    (b) the court by or before which he is so convicted is satisfied that the health risk condition is fulfilled with respect to that business,

the court shall by an order impose the appropriate prohibition.

(2) The health risk condition is fulfilled with respect to any food business if any of the following involves risk of injury to health, namely—

    (a) the use for the purposes of the business of any process or treatment;

    (b) the construction of any premises used for the purposes of the business, or the use for those purposes of any equipment; and

(c) the state or condition of any premises or equipment used for the purposes of the business.

(3) The appropriate prohibition is—

(a) in a case falling within paragraph (a) of subsection (2) above, a prohibition on the use of the process or treatment for the purposes of the business;

(b) in a case falling within paragraph (b) of that subsection, a prohibition on the use of the premises or equipment for the purposes of the business or any other food business of the same class or description;

(c) in a case falling within paragraph (c) of that subsection, a prohibition on the use of the premises or equipment for the purposes of any food business.

(4) If—

(a) the proprietor of a food business is convicted of an offence under any regulations to which this section applies by virtue of section 10(3)(b) above; and

(b) the court by or before which he is so convicted thinks it proper to do so in all the circumstances of the case,

the court may, by an order, impose a prohibition on the proprietor participating in the management of any food business, or any food business of a class or description specified in the order.

(5) As soon as practicable after the making of an order under subsection (1) or (4) above (in this Act referred to as a "prohibition order"), the enforcement authority shall—

(a) serve a copy of the order on the proprietor of the business; and

(b) in the case of an order under subsection (1) above, affix a copy of the order in a conspicuous position on such premises used for the purposes of the business as they consider appropriate;

and any person who knowingly contravenes such an order shall be guilty of an offence.

(6) A prohibition order shall cease to have effect—

(a) in the case of an order under subsection (1) above, on the issue by the enforcement authority of a certificate to the effect that they are satisfied that the proprietor has taken sufficient measures to secure that the health risk condition is no longer fulfilled with respect to the business;

(b) in the case of an order under subsection (4) above, on the giving by the court of a direction to that effect.

(7) The enforcement authority shall issue a certificate under paragraph (a) of subsection (6) above within three days of their being satisfied as mentioned in that paragraph; and on an application by the proprietor for such a certificate, the authority shall—

(a) determine, as soon as is reasonably practicable and in any event within 14 days, whether or not they are so satisfied; and

(b) if they determine that they are not so satisfied, give notice to the proprietor of the reasons for that determination.

(8) The court shall give a direction under subsection (6)(b) above if, on an application by the proprietor, the court thinks it proper to do so having regard to all the circumstances of the case, including in particular the conduct of the proprietor since the making of the order; but no such application shall be entertained if it is made—

    (a) within six months after the making of the prohibition order; or

    (b) within three months after the making by the proprietor of a previous application for such a direction.

(9) Where a magistrates' court or, in Scotland, the sheriff makes an order under section 12(2) below with respect to any food business, subsection (1) above shall apply as if the proprietor of the business had been convicted by the court or sheriff of an offence under regulations to which this section applies.

(10) Subsection (4) above shall apply in relation to a manager of a food business as it applies in relation to the proprietor of such a business; and any reference in subsection (5) or (8) above to the proprietor of the business, or to the proprietor, shall be construed accordingly.

(11) In subsection (10) above "manager", in relation to a food business, means any person who is entrusted by the proprietor with the day to day running of the business, or any part of the business.

Emergency prohibition notices and orders.

12.—(1) If an authorised officer of an enforcement authority is satisfied that the health risk condition is fulfilled with respect to any food business, he may, by a notice served on the proprietor of the business (in this Act referred to as an "emergency prohibition notice"), impose the appropriate prohibition.

(2) If a magistrates' court or, in Scotland, the sheriff is satisfied, on the application of such an officer, that the health risk condition is fulfilled with respect to any food business, the court or sheriff shall, by an order (in this Act referred to as an "emergency prohibition order"), impose the appropriate prohibition.

(3) Such an officer shall not apply for an emergency prohibition order unless, at least one day before the date of the application, he has served notice on the proprietor of the business of his intention to apply for the order.

(4) Subsections (2) and (3) of section 11 above shall apply for the purposes of this section as they apply for the purposes of that section, but as if the reference in subsection (2) to risk of injury to health were a reference to imminent risk of such injury.

(5) As soon as practicable after the service of an emergency prohibition notice, the enforcement authority shall affix a copy of the notice in a conspicuous position on such premises used for the purposes of the business as they consider appropriate; and any person who knowingly contravenes such a notice shall be guilty of an offence.

(6) As soon as practicable after the making of an emergency prohibition order, the enforcement authority shall—

    (a) serve a copy of the order on the proprietor of the business; and

(b) affix a copy of the order in a conspicuous position on such premises used for the purposes of that business as they consider appropriate;

and any person who knowingly contravenes such an order shall be guilty of an offence.

(7) An emergency prohibition notice shall cease to have effect—

(a) if no application for an emergency prohibition order is made within the period of three days beginning with the service of the notice, at the end of that period;

(b) if such an application is so made, on the determination or abandonment of the application.

(8) An emergency prohibition notice or emergency prohibition order shall cease to have effect on the issue by the enforcement authority of a certificate to the effect that they are satisfied that the proprietor has taken sufficient measures to secure that the health risk condition is no longer fulfilled with respect to the business.

(9) The enforcement authority shall issue a certificate under subsection (8) above within three days of their being satisfied as mentioned in that subsection; and on an application by the proprietor for such a certificate, the authority shall—

(a) determine, as soon as is reasonably practicable and in any event within 14 days, whether or not they are so satisfied; and

(b) if they determine that they are not so satisfied, give notice to the proprietor of the reasons for that determination.

(10) Where an emergency prohibition notice is served on the proprietor of a business, the enforcement authority shall compensate him in respect of any loss suffered by reason of his complying with the notice unless—

(a) an application for an emergency prohibition order is made within the period of three days beginning with the service of the notice; and

(b) the court declares itself satisfied, on the hearing of the application, that the health risk condition was fulfilled with respect to the business at the time when the notice was served;

and any disputed question as to the right to or the amount of any compensation payable under this subsection shall be determined by arbitration or, in Scotland, by a single arbiter appointed, failing agreement between the parties, by the sheriff.

13.—(1) If it appears to the Minister that the carrying out of commercial operations with respect to food, food sources or contact materials of any class or description involves or may involve imminent risk of injury to health, he may, by an order (in this Act referred to as an "emergency control order"), prohibit the carrying out of such operations with respect to food, food sources or contact materials of that class or description.

Emergency control orders.

(2) Any person who knowingly contravenes an emergency control order shall be guilty of an offence.

(3) The Minister may consent, either unconditionally or subject to any condition that he considers appropriate, to the doing in a particular case of anything prohibited by an emergency control order.

(4) It shall be a defence for a person charged with an offence under subsection (2) above to show—

(a) that consent had been given under subsection (3) above to the contravention of the emergency control order; and

(b) that any condition subject to which that consent was given was complied with.

(5) The Minister—

(a) may give such directions as appear to him to be necessary or expedient for the purpose of preventing the carrying out of commercial operations with respect to any food, food sources or contact materials which he believes, on reasonable grounds, to be food, food sources or contact materials to which an emergency control order applies; and

(b) may do anything which appears to him to be necessary or expedient for that purpose.

(6) Any person who fails to comply with a direction under this section shall be guilty of an offence.

(7) If the Minister does anything by virtue of this section in consequence of any person failing to comply with an emergency control order or a direction under this section, the Minister may recover from that person any expenses reasonably incurred by him under this section.

Consumer protection

Selling food not of the nature or substance or quality demanded.

14.—(1) Any person who sells to the purchaser's prejudice any food which is not of the nature or substance or quality demanded by the purchaser shall be guilty of an offence.

(2) In subsection (1) above the reference to sale shall be construed as a reference to sale for human consumption; and in proceedings under that subsection it shall not be a defence that the purchaser was not prejudiced because he bought for analysis or examination.

Falsely describing or presenting food.

15.—(1) Any person who gives with any food sold by him, or displays with any food offered or exposed by him for sale or in his possession for the purpose of sale, a label, whether or not attached to or printed on the wrapper or container, which—

(a) falsely describes the food; or

(b) is likely to mislead as to the nature or substance or quality of the food,

shall be guilty of an offence.

(2) Any person who publishes, or is a party to the publication of, an advertisement (not being such a label given or displayed by him as mentioned in subsection (1) above) which—

(a) falsely describes any food; or

(b) is likely to mislead as to the nature or substance or quality of any food,

shall be guilty of an offence.

(3) Any person who sells, or offers or exposes for sale, or has in his possession for the purpose of sale, any food the presentation of which is likely to mislead as to the nature or substance or quality of the food shall be guilty of an offence.

(4) In proceedings for an offence under subsection (1) or (2) above, the fact that a label or advertisement in respect of which the offence is alleged to have been committed contained an accurate statement of the composition of the food shall not preclude the court from finding that the offence was committed.

(5) In this section references to sale shall be construed as references to sale for human consumption.

Regulations

16.—(1) The Ministers may by regulations make—

    (a) provision for requiring, prohibiting or regulating the presence in food or food sources of any specified substance, or any substance of any specified class, and generally for regulating the composition of food;

    (b) provision for securing that food is fit for human consumption and meets such microbiological standards (whether going to the fitness of the food or otherwise) as may be specified by or under the regulations;

    (c) provision for requiring, prohibiting or regulating the use of any process or treatment in the preparation of food;

    (d) provision for securing the observance of hygienic conditions and practices in connection with the carrying out of commercial operations with respect to food or food sources;

    (e) provision for imposing requirements or prohibitions as to, or otherwise regulating, the labelling, marking, presenting or advertising of food, and the descriptions which may be applied to food; and

    (f) such other provision with respect to food or food sources, including in particular provision for prohibiting or regulating the carrying out of commercial operations with respect to food or food sources, as appears to them to be necessary or expedient—

        (i) for the purpose of securing that food complies with food safety requirements or in the interests of the public health; or

        (ii) for the purpose of protecting or promoting the interests of consumers.

(2) The Ministers may also by regulations make provision—

    (a) for securing the observance of hygienic conditions and practices in connection with the carrying out of commercial operations with respect to contact materials which are intended to come into contact with food intended for human consumption;

    (b) for imposing requirements or prohibitions as to, or otherwise regulating, the labelling, marking or advertising of such materials, and the descriptions which may be applied to them; and

PART II

(c) otherwise for prohibiting or regulating the carrying out of commercial operations with respect to such materials.

(3) Without prejudice to the generality of subsection (1) above, regulations under that subsection may make any such provision as is mentioned in Schedule 1 to this Act.

(4) In making regulations under subsection (1) above, the Ministers shall have regard to the desirability of restricting, so far as practicable, the use of substances of no nutritional value as foods or as ingredients of foods.

(5) In subsection (1) above and Schedule 1 to this Act, unless the context otherwise requires—

(a) references to food shall be construed as references to food intended for sale for human consumption; and

(b) references to food sources shall be construed as references to food sources from which such food is intended to be derived.

Enforcement of Community provisions.

17.—(1) The Ministers may by regulations make such provision with respect to food, food sources or contact materials, including in particular provision for prohibiting or regulating the carrying out of commercial operations with respect to food, food sources or contact materials, as appears to them to be called for by any Community obligation.

(2) As respects any directly applicable Community provision which relates to food, food sources or contact materials and for which, in their opinion, it is appropriate to provide under this Act, the Ministers may by regulations—

(a) make such provision as they consider necessary or expedient for the purpose of securing that the Community provision is administered, executed and enforced under this Act; and

(b) apply such of the provisions of this Act as may be specified in the regulations in relation to the Community provision with such modifications, if any, as may be so specified.

(3) In subsections (1) and (2) above references to food or food sources shall be construed in accordance with section 16(5) above.

Special provisions for particular foods etc.

18.—(1) The Ministers may by regulations make provision—

(a) for prohibiting the carrying out of commercial operations with respect to novel foods, or food sources from which such foods are intended to be derived, of any class specified in the regulations;

(b) for prohibiting the carrying out of such operations with respect to genetically modified food sources, or foods derived from such food sources, of any class so specified; or

(c) for prohibiting the importation of any food of a class so specified,

and (in each case) for excluding from the prohibition any food or food source which is of a description specified by or under the regulations and, in the case of a prohibition on importation, is imported at an authorised place of entry.

(2) The Ministers may also by regulations—

(a) prescribe, in relation to milk of any description, such a designation (in this subsection referred to as a "special designation") as the Ministers consider appropriate;

(b) provide for the issue by enforcement authorities of licences to producers and sellers of milk authorising the use of a special designation; and

(c) prohibit, without the use of a special designation, all sales of milk for human consumption, other than sales made with the Minister's consent.

(3) In this section—

"authorised place of entry" means any port, aerodrome or other place of entry authorised by or under the regulations and, in relation to food in a particular consignment, includes any place of entry so authorised for the importation of that consignment;

"description", in relation to food, includes any description of its origin or of the manner in which it is packed;

"novel food" means any food which has not previously been used for human consumption in Great Britain, or has been so used only to a very limited extent.

(4) For the purposes of this section a food source is genetically modified if any of the genes or other genetic material in the food source—

(a) has been modified by means of an artificial technique; or

(b) is inherited or otherwise derived, through any number of replications, from genetic material which was so modified;

and in this subsection "artificial technique" does not include any technique which involves no more than, or no more than the assistance of, naturally occurring processes of reproduction (including selective breeding techniques or *in vitro* fertilisation).

19.—(1) The Ministers may by regulations make provision—

Registration and licensing of food premises.

(a) for the registration by enforcement authorities of premises used or proposed to be used for the purposes of a food business, and for prohibiting the use for those purposes of any premises which are not registered in accordance with the regulations; or

(b) subject to subsection (2) below, for the issue by such authorities of licences in respect of the use of premises for the purposes of a food business, and for prohibiting the use for those purposes of any premises except in accordance with a licence issued under the regulations.

(2) The Ministers shall exercise the power conferred by subsection (1)(b) above only where it appears to them to be necessary or expedient to do so—

(a) for the purpose of securing that food complies with food safety requirements or in the interests of the public health; or

(b) for the purpose of protecting or promoting the interests of consumers.

Defences etc.

Offences due to
fault of another
person.

20. Where the commission by any person of an offence under any of the preceding provisions of this Part is due to an act or default of some other person, that other person shall be guilty of the offence; and a person may be charged with and convicted of the offence by virtue of this section whether or not proceedings are taken against the first-mentioned person.

Defence of due
diligence.

21.—(1) In any proceedings for an offence under any of the preceding provisions of this Part (in this section referred to as "the relevant provision"), it shall, subject to subsection (5) below, be a defence for the person charged to prove that he took all reasonable precautions and exercised all due diligence to avoid the commission of the offence by himself or by a person under his control.

(2) Without prejudice to the generality of subsection (1) above, a person charged with an offence under section 8, 14 or 15 above who neither—

(a) prepared the food in respect of which the offence is alleged to have been committed; nor

(b) imported it into Great Britain,

shall be taken to have established the defence provided by that subsection if he satisfies the requirements of subsection (3) or (4) below.

(3) A person satisfies the requirements of this subsection if he proves—

(a) that the commission of the offence was due to an act or default of another person who was not under his control, or to reliance on information supplied by such a person;

(b) that he carried out all such checks of the food in question as were reasonable in all the circumstances, or that it was reasonable in all the circumstances for him to rely on checks carried out by the person who supplied the food to him; and

(c) that he did not know and had no reason to suspect at the time of the commission of the alleged offence that his act or omission would amount to an offence under the relevant provision.

(4) A person satisfies the requirements of this subsection if he proves—

(a) that the commission of the offence was due to an act or default of another person who was not under his control, or to reliance on information supplied by such a person;

(b) that the sale or intended sale of which the alleged offence consisted was not a sale or intended sale under his name or mark; and

(c) that he did not know, and could not reasonably have been expected to know, at the time of the commission of the alleged offence that his act or omission would amount to an offence under the relevant provision.

(5) If in any case the defence provided by subsection (1) above involves the allegation that the commission of the offence was due to an act or default of another person, or to reliance on information supplied by another person, the person charged shall not, without leave of the court, be entitled to rely on that defence unless—

(a) at least seven clear days before the hearing; and

(b) where he has previously appeared before a court in connection with the alleged offence, within one month of his first such appearance,

he has served on the prosecutor a notice in writing giving such information identifying or assisting in the identification of that other person as was then in his possession.

(6) In subsection (5) above any reference to appearing before a court shall be construed as including a reference to being brought before a court.

---

22. In proceedings for an offence under any of the preceding provisions of this Part consisting of the advertisement for sale of any food, it shall be a defence for the person charged to prove—

Defence of publication in the course of business.

    (a) that he is a person whose business it is to publish or arrange for the publication of advertisements; and

    (b) that he received the advertisement in the ordinary course of business and did not know and had no reason to suspect that its publication would amount to an offence under that provision.

Miscellaneous and supplemental

23.—(1) A food authority may provide, whether within or outside their area, training courses in food hygiene for persons who are or intend to become involved in food businesses, whether as proprietors or employees or otherwise.

Provision of food hygiene training.

(2) A food authority may contribute towards the expenses incurred under this section by any other such authority, or towards expenses incurred by any other person in providing, such courses as are mentioned in subsection (1) above.

24.—(1) A food authority may provide, whether within or outside their area, tanks or other apparatus for cleansing shellfish.

Provision of facilities for cleansing shellfish.

(2) A food authority may contribute towards the expenses incurred under this section by any other such authority, or towards expenses incurred by any other person in providing, and making available to the public, tanks or other apparatus for cleansing shellfish.

(3) Nothing in this section authorises the establishment of any tank or other apparatus, or the execution of any other work, on, over or under tidal lands below high-water mark of ordinary spring tides, except in accordance with such plans and sections, and subject to such restrictions and conditions as may before the work is commenced be approved by the Secretary of State.

(4) In this section "cleansing", in relation to shellfish, includes subjecting them to any germicidal treatment.

25.—(1) For the purpose of facilitating the exercise of their functions under this Part, the Ministers may by order require every person who at the date of the order, or at any subsequent time, carries on a business of a specified class or description (in this section referred to as a "relevant business")—

Orders for facilitating the exercise of functions.

(a) to afford to persons specified in the order such facilities for the taking of samples of any food, substance or contact material to which subsection (2) below applies; or

(b) to furnish to persons so specified such information concerning any such food, substance or contact material,

as (in each case) is specified in the order and is reasonably required by such persons.

(2) This subsection applies to—

(a) any food of a class specified in the order which is sold or intended to be sold in the course of a relevant business for human consumption;

(b) any substance of a class so specified which is sold in the course of such a business for use in the preparation of food for human consumption, or is used for that purpose in the course of such a business; and

(c) any contact material of a class so specified which is sold in the course of such a business and is intended to come into contact with food intended for human consumption.

(3) No information relating to any individual business which is obtained by means of an order under subsection (1) above shall, without the previous consent in writing of the person carrying on the business, be disclosed except—

(a) in accordance with directions of the Minister, so far as may be necessary for the purposes of this Act or of any corresponding enactment in force in Northern Ireland, or for the purpose of complying with any Community obligation; or

(b) for the purposes of any proceedings for an offence against the order or any report of those proceedings;

and any person who discloses any such information in contravention of this subsection shall be guilty of an offence.

(4) In subsection (3) above the reference to a disclosure being necessary for the purposes of this Act includes a reference to it being necessary—

(a) for the purpose of securing that food complies with food safety requirements or in the interests of the public health; or

(b) for the purpose of protecting or promoting the interests of consumers;

and the reference to a disclosure being necessary for the purposes of any corresponding enactment in force in Northern Ireland shall be construed accordingly.

Regulations and orders: supplementary provisions.

26.—(1) Regulations under this Part may—

(a) make provision for prohibiting or regulating the carrying out of commercial operations with respect to any food, food source or contact material—

(i) which fails to comply with the regulations; or

(ii) in relation to which an offence against the regulations has been committed, or would have been committed if any relevant act or omission had taken place in Great Britain; and

(b) without prejudice to the generality of section 9 above, provide that any food which, in accordance with the regulations, is certified as being such food as is mentioned in paragraph (a) above may be treated for the purposes of that section as failing to comply with food safety requirements.

(2) Regulations under this Part may also—

    (a) require persons carrying on any activity to which the regulations apply to keep and produce records and provide returns;

    (b) prescribe the particulars to be entered on any register required to be kept in accordance with the regulations;

    (c) require any such register to be open to inspection by the public at all reasonable times and, subject to that, authorise it to be kept by means of a computer;

    (d) prescribe the periods for which and the conditions subject to which licences may be issued, and provide for the subsequent alteration of conditions and for the cancellation, suspension or revocation of licences;

    (e) provide for an appeal to a magistrates' court or, in Scotland, to the sheriff, or to a tribunal constituted in accordance with the regulations, against any decision of an enforcement authority, or of an authorised officer of such an authority; and

    (f) provide, as respects any appeal to such a tribunal, for the procedure on the appeal (including costs) and for any appeal against the tribunal's decision.

(3) Regulations under this Part or an order under section 25 above may—

    (a) provide that an offence under the regulations or order shall be triable in such way as may be there specified; and

    (b) include provisions under which a person guilty of such an offence shall be liable to such penalties (not exceeding those which may be imposed in respect of offences under this Act) as may be specified in the regulations or order.

PART III

ADMINISTRATION AND ENFORCEMENT

Administration

27.—(1) Every authority to whom this section applies, that is to say, every food authority in England and Wales and every regional or islands council in Scotland, shall appoint in accordance with this section one or more persons (in this Act referred to as "public analysts") to act as analysts for the purposes of this Act within the authority's area.

Appointment of public analysts.

(2) No person shall be appointed as a public analyst unless he possesses—

    (a) such qualifications as may be prescribed by regulations made by the Ministers; or

    (b) such other qualifications as the Ministers may approve,

and no person shall act as a public analyst for any area who is engaged directly or indirectly in any food business which is carried on in that area.

(3) An authority to whom this section applies shall pay to a public analyst such remuneration as may be agreed, which may be expressed to be payable either—

   (a) in addition to any fees received by him under this Part; or

   (b) on condition that any fees so received by him are paid over by him to the authority.

(4) An authority to whom this section applies who appoint only one public analyst may appoint also a deputy to act during any vacancy in the office of public analyst, or during the absence or incapacity of the holder of the office, and—

   (a) the provisions of this section with respect to the qualifications, appointment, removal and remuneration of a public analyst shall apply also in relation to a deputy public analyst; and

   (b) any reference in the following provisions of this Act to a public analyst shall be construed as including a reference to a deputy public analyst appointed under this subsection.

(5) In subsection (1) above "food authority" does not include the council of a non-metropolitan district, the Sub-Treasurer of the Inner Temple or the Under Treasurer of the Middle Temple; and in subsection (2) above the reference to being engaged directly or indirectly in a food business includes a reference to having made such arrangements with a food business as may be prescribed by regulations made by the Ministers.

Provision of facilities for examinations.

28.—(1) A food authority, or a regional council in Scotland, may provide facilities for examinations for the purposes of this Act.

(2) In this Act "examination" means a microbiological examination and "examine" shall be construed accordingly.

Sampling and analysis etc.

Procurement of samples.

29. An authorised officer of an enforcement authority may—

   (a) purchase a sample of any food, or any substance capable of being used in the preparation of food;

   (b) take a sample of any food, or any such substance, which—

      (i) appears to him to be intended for sale, or to have been sold, for human consumption; or

      (ii) is found by him on or in any premises which he is authorised to enter by or under section 32 below;

   (c) take a sample from any food source, or a sample of any contact material, which is found by him on or in any such premises;

   (d) take a sample of any article or substance which is found by him on or in any such premises and which he has reason to believe may be required as evidence in proceedings under any of the provisions of this Act or of regulations or orders made under it.

Analysis etc. of samples.

30.—(1) An authorised officer of an enforcement authority who has procured a sample under section 29 above shall—

   (a) if he considers that the sample should be analysed, submit it to be analysed either—

(i) by the public analyst for the area in which the sample was procured; or

(ii) by the public analyst for the area which consists of or includes the area of the authority;

(b) if he considers that the sample should be examined, submit it to be examined by a food examiner.

(2) A person, other than such an officer, who has purchased any food, or any substance capable of being used in the preparation of food, may submit a sample of it—

(a) to be analysed by the public analyst for the area in which the purchase was made; or

(b) to be examined by a food examiner.

(3) If, in any case where a sample is proposed to be submitted for analysis under this section, the office of public analyst for the area in question is vacant, the sample shall be submitted to the public analyst for some other area.

(4) If, in any case where a sample is proposed to be or is submitted for analysis or examination under this section, the food analyst or examiner determines that he is for any reason unable to perform the analysis or examination, the sample shall be submitted or, as the case may be, sent by him to such other food analyst or examiner as he may determine.

(5) A food analyst or examiner shall analyse or examine as soon as practicable any sample submitted or sent to him under this section, but may, except where—

(a) he is the public analyst for the area in question; and

(b) the sample is submitted to him for analysis by an authorised officer of an enforcement authority,

demand in advance the payment of such reasonable fee as he may require.

(6) A food analyst or examiner who has analysed or examined a sample shall give to the person by whom it was submitted a certificate specifying the result of the analysis or examination.

(7) Any certificate given by a food analyst or examiner under subsection (6) above shall be signed by him, but the analysis or examination may be made by any person acting under his direction.

(8) In any proceedings under this Act, the production by one of the parties—

(a) of a document purporting to be a certificate given by a food analyst or examiner under subsection (6) above; or

(b) of a document supplied to him by the other party as being a copy of such a certificate,

shall be sufficient evidence of the facts stated in it unless, in a case falling within paragraph (a) above, the other party requires that the food analyst or examiner shall be called as a witness.

(9) In this section—

"food analyst" means a public analyst or any other person who possesses the requisite qualifications to carry out analyses for the purposes of this Act;

"food examiner" means any person who possesses the requisite qualifications to carry out examinations for the purposes of this Act;

"the requisite qualifications" means such qualifications as may be prescribed by regulations made by the Ministers, or such other qualifications as the Ministers may approve;

"sample", in relation to an authorised officer of an enforcement authority, includes any part of a sample retained by him in pursuance of regulations under section 31 below;

and where two or more public analysts are appointed for any area, any reference in this section to the public analyst for that area shall be construed as a reference to either or any of them.

Regulation of sampling and analysis etc.

31.—(1) The Ministers may by regulations make provision for supplementing or modifying the provisions of sections 29 and 30 above.

(2) Without prejudice to the generality of subsection (1) above, regulations under that subsection may make provision with respect to—

(a) the matters to be taken into account in determining whether, and at what times, samples should be procured;

(b) the manner of procuring samples, including the steps to be taken in order to ensure that any samples procured are fair samples;

(c) the method of dealing with samples, including (where appropriate) their division into parts;

(d) the persons to whom parts of samples are to be given and the persons by whom such parts are to be retained;

(e) the notices which are to be given to, and the information which is to be furnished by, the persons in charge of any food, substance, contact material or food source of or from which samples are procured;

(f) the methods which are to be used in analysing or examining samples, or parts of samples, or in classifying the results of analyses or examinations;

(g) the circumstances in which a food analyst or examiner is to be precluded, by reason of a conflict of interest, from analysing or examining a particular sample or part of a sample; and

(h) the circumstances in which samples, or parts of samples, are to be or may be submitted for analysis or examination—

(i) to the Government Chemist, or to such other food analyst or examiner as he may direct; or

(ii) to a person determined by or under the regulations.

(3) In this section "food analyst" and "food examiner" have the same meanings as in section 30 above.

Powers of entry and obstruction etc.

Powers of entry.

32.—(1) An authorised officer of an enforcement authority shall, on producing, if so required, some duly authenticated document showing his authority, have a right at all reasonable hours—

(a) to enter any premises within the authority's area for the purpose of ascertaining whether there is or has been on the premises any contravention of the provisions of this Act, or of regulations or orders made under it; and

(b) to enter any business premises, whether within or outside the authority's area, for the purpose of ascertaining whether there is on the premises any evidence of any contravention within that area of any of such provisions; and

(c) in the case of an authorised officer of a food authority, to enter any premises for the purpose of the performance by the authority of their functions under this Act;

but admission to any premises used only as a private dwelling-house shall not be demanded as of right unless 24 hours' notice of the intended entry has been given to the occupier.

(2) If a justice of the peace, on sworn information in writing, is satisfied that there is reasonable ground for entry into any premises for any such purpose as is mentioned in subsection (1) above and either—

(a) that admission to the premises has been refused, or a refusal is apprehended, and that notice of the intention to apply for a warrant has been given to the occupier; or

(b) that an application for admission, or the giving of such a notice, would defeat the object of the entry, or that the case is one of urgency, or that the premises are unoccupied or the occupier temporarily absent,

the justice may by warrant signed by him authorise the authorised officer to enter the premises, if need be by reasonable force.

(3) Every warrant granted under this section shall continue in force for a period of one month.

(4) An authorised officer entering any premises by virtue of this section, or of a warrant issued under it, may take with him such other persons as he considers necessary, and on leaving any unoccupied premises which he has entered by virtue of such a warrant shall leave them as effectively secured against unauthorised entry as he found them.

(5) An authorised officer entering premises by virtue of this section, or of a warrant issued under it, may inspect any records (in whatever form they are held) relating to a food business and, where any such records are kept by means of a computer—

(a) may have access to, and inspect and check the operation of, any computer and any associated apparatus or material which is or has been in use in connection with the records; and

(b) may require any person having charge of, or otherwise concerned with the operation of, the computer, apparatus or material to afford him such assistance as he may reasonably require.

(6) Any officer exercising any power conferred by subsection (5) above may—

(a) seize and detain any records which he has reason to believe may be required as evidence in proceedings under any of the provisions of this Act or of regulations or orders made under it; and

(b) where the records are kept by means of a computer, may require the records to be produced in a form in which they may be taken away.

(7) If any person who enters any premises by virtue of this section, or of a warrant issued under it, discloses to any person any information obtained by him in the premises with regard to any trade secret, he shall, unless the disclosure was made in the performance of his duty, be guilty of an offence.

1981 c. 22.
(8) Nothing in this section authorises any person, except with the permission of the local authority under the Animal Health Act 1981, to enter any premises—

(a) in which an animal or bird affected with any disease to which that Act applies is kept; and

(b) which is situated in a place declared under that Act to be infected with such a disease.

(9) In the application of this section to Scotland, any reference to a justice of the peace includes a reference to the sheriff and to a magistrate.

Obstruction etc. of officers.
33.—(1) Any person who—

(a) intentionally obstructs any person acting in the execution of this Act; or

(b) without reasonable cause, fails to give to any person acting in the execution of this Act any assistance or information which that person may reasonably require of him for the performance of his functions under this Act,

shall be guilty of an offence.

(2) Any person who, in purported compliance with any such requirement as is mentioned in subsection (1)(b) above—

(a) furnishes information which he knows to be false or misleading in a material particular; or

(b) recklessly furnishes information which is false or misleading in a material particular,

shall be guilty of an offence.

(3) Nothing in subsection (1)(b) above shall be construed as requiring any person to answer any question or give any information if to do so might incriminate him.

Offences

Time limit for prosecutions.
34. No prosecution for an offence under this Act which is punishable under section 35(2) below shall be begun after the expiry of—

(a) three years from the commission of the offence; or

(b) one year from its discovery by the prosecutor,

whichever is the earlier.

Punishment of offences.
35.—(1) A person guilty of an offence under section 33(1) above shall be liable on summary conviction to a fine not exceeding level 5 on the standard scale or to imprisonment for a term not exceeding three months or to both.

(2) A person guilty of any other offence under this Act shall be liable—

   (a) on conviction on indictment, to a fine or to imprisonment for a term not exceeding two years or to both;

   (b) on summary conviction, to a fine not exceeding the relevant amount or to imprisonment for a term not exceeding six months or to both.

(3) In subsection (2) above "the relevant amount" means—

   (a) in the case of an offence under section 7, 8 or 14 above, £20,000;

   (b) in any other case, the statutory maximum.

(4) If a person who is—

   (a) licensed under section 1 of the Slaughterhouses Act 1974 to keep a slaughterhouse or knacker's yard;   *1974 c. 3.*

   (b) registered under section 4 of the Slaughter of Animals (Scotland) Act 1980 in respect of any premises for use as a slaughterhouse; or   *1980 c. 13.*

   (c) licensed under section 6 of that Act to use any premises as a knacker's yard,

is convicted of an offence under Part II of this Act, the court may, in addition to any other punishment, cancel his licence or registration.

36.—(1) Where an offence under this Act which has been committed by a body corporate is proved to have been committed with the consent or connivance of, or to be attributable to any neglect on the part of—   Offences by bodies corporate.

   (a) any director, manager, secretary or other similar officer of the body corporate; or

   (b) any person who was purporting to act in any such capacity,

he as well as the body corporate shall be deemed to be guilty of that offence and shall be liable to be proceeded against and punished accordingly.

(2) In subsection (1) above "director", in relation to any body corporate established by or under any enactment for the purpose of carrying on under national ownership any industry or part of an industry or undertaking, being a body corporate whose affairs are managed by its members, means a member of that body corporate.

Appeals

37.—(1) Any person who is aggrieved by—   Appeals to magistrates' court or sheriff.

   (a) a decision of an authorised officer of an enforcement authority to serve an improvement notice;

   (b) a decision of an enforcement authority to refuse to issue such a certificate as is mentioned in section 11(6) or 12(8) above; or

   (c) subject to subsection (2) below, a decision of such an authority to refuse, cancel, suspend or revoke a licence required by regulations under Part II of this Act,

may appeal to a magistrates' court or, in Scotland, to the sheriff.

(2) Subsection (1)(c) above shall not apply in relation to any decision as respects which regulations under Part II of this Act provide for an appeal to a tribunal constituted in accordance with the regulations.

PART III

1980 c. 43.

(3) The procedure on an appeal to a magistrates' court under subsection (1) above, or an appeal to such a court for which provision is made by regulations under Part II of this Act, shall be by way of complaint for an order, and the Magistrates' Courts Act 1980 shall apply to the proceedings.

(4) An appeal to the sheriff under subsection (1) above, or an appeal to the sheriff for which provision is made by regulations under Part II of this Act, shall be by summary application.

(5) The period within which such an appeal as is mentioned in subsection (3) or (4) above may be brought shall be—

(a) one month from the date on which notice of the decision was served on the person desiring to appeal; or

(b) in the case of an appeal under subsection (1)(a) above, that period or the period specified in the improvement notice, whichever ends the earlier;

and, in the case of such an appeal as is mentioned in subsection (3) above, the making of the complaint shall be deemed for the purposes of this subsection to be the bringing of the appeal.

(6) In any case where such an appeal as is mentioned in subsection (3) or (4) above lies, the document notifying the decision to the person concerned shall state—

(a) the right of appeal to a magistrates' court or to the sheriff; and

(b) the period within which such an appeal may be brought.

Appeals to Crown Court.

38. A person who is aggrieved by—

(a) any dismissal by a magistrates' court of such an appeal as is mentioned in section 37(3) above; or

(b) any decision of such a court to make a prohibition order or an emergency prohibition order, or to exercise the power conferred by section 35(4) above,

may appeal to the Crown Court.

Appeals against improvement notices.

39.—(1) On an appeal against an improvement notice, the court may either cancel or affirm the notice and, if it affirms it, may do so either in its original form or with such modifications as the court may in the circumstances think fit.

(2) Where, apart from this subsection, any period specified in an improvement notice would include any day on which an appeal against that notice is pending, that day shall be excluded from that period.

(3) An appeal shall be regarded as pending for the purposes of subsection (2) above until it is finally disposed of, is withdrawn or is struck out for want of prosecution.

PART IV

MISCELLANEOUS AND SUPPLEMENTAL

Powers of Ministers

40.—(1) For the guidance of food authorities, the Ministers or the Minister may issue codes of recommended practice as regards the execution and enforcement of this Act and of regulations and orders made under it; and any such code shall be laid before Parliament after being issued.

Power to issue codes of practice.

(2) In the exercise of the functions conferred on them by or under this Act, every food authority—

    (a) shall have regard to any relevant provision of any such code; and

    (b) shall comply with any direction which is given by the Ministers or the Minister and requires them to take any specified steps in order to comply with such a code.

(3) Any direction under subsection (2)(b) above shall, on the application of the Ministers or the Minister, be enforceable by mandamus or, in Scotland, by an order of the Court of Session under section 45 of the Court of Session Act 1988.

1988 c. 36.

(4) Before issuing any code under this section, the Ministers or the Minister shall consult with such organisations as appear to them or him to be representative of interests likely to be substantially affected by the code.

(5) Any consultation undertaken before the commencement of subsection (4) above shall be as effective, for the purposes of that subsection, as if undertaken after that commencement.

41. Every food authority shall send to the Minister such reports and returns, and give him such information, with respect to the exercise of the functions conferred on them by or under this Act as he may require.

Power to require returns.

42.—(1) Where the Minister is satisfied that—

Default powers.

    (a) a food authority (in this section referred to as "the authority in default") have failed to discharge any duty imposed by or under this Act; and

    (b) the authority's failure affects the general interests of consumers of food,

he may by order empower another food authority (in this section referred to as "the substitute authority"), or one of his officers, to discharge that duty in place of the authority in default.

(2) For the purpose of determining whether the power conferred by subsection (1) above is exercisable, the Minister may cause a local inquiry to be held; and where he does so, the relevant provisions of the Local Government Act shall apply as if the inquiry were a local inquiry held under that Act.

(3) Nothing in subsection (1) above affects any other power exercisable by the Minister with respect to defaults of local authorities.

(4) The substitute authority or the Minister may recover from the authority in default any expenses reasonably incurred by them or him under subsection (1) above; and for the purpose of paying any such amount the authority in default may—

  (a) raise money as if the expenses had been incurred directly by them as a local authority; and

  (b) if and to the extent that they are authorised to do so by the Minister, borrow money in accordance with the statutory provisions relating to borrowing by a local authority.

(5) In this section "the relevant provisions of the Local Government Act" means subsections (2) to (5) of section 250 of the Local Government Act 1972 in relation to England and Wales and subsections (3) to (8) of section 210 of the Local Government (Scotland) Act 1973 in relation to Scotland.

1972 c. 70.

1973 c. 65.

Protective provisions

Continuance of registration or licence on death.

43.—(1) This section shall have effect on the death of any person who—

  (a) is registered in respect of any premises in accordance with regulations made under Part II of this Act; or

  (b) holds a licence issued in accordance with regulations so made.

(2) The registration or licence shall subsist for the benefit of the deceased's personal representative, or his widow or any other member of his family, until the end of—

  (a) the period of three months beginning with his death; or

  (b) such longer period as the enforcement authority may allow.

Protection of officers acting in good faith.

44.—(1) An officer of a food authority is not personally liable in respect of any act done by him—

  (a) in the execution or purported execution of this Act; and

  (b) within the scope of his employment,

if he did that act in the honest belief that his duty under this Act required or entitled him to do it.

(2) Nothing in subsection (1) above shall be construed as relieving any food authority from any liability in respect of the acts of their officers.

(3) Where an action has been brought against an officer of a food authority in respect of an act done by him—

  (a) in the execution or purported execution of this Act; but

  (b) outside the scope of his employment,

the authority may indemnify him against the whole or a part of any damages which he has been ordered to pay or any costs which he may have incurred if they are satisfied that he honestly believed that the act complained of was within the scope of his employment.

(4) A public analyst appointed by a food authority shall be treated for the purposes of this section as being an officer of the authority, whether or not his appointment is a whole-time appointment.

Financial provisions

45.—(1) The Ministers may make regulations requiring or authorising charges to be imposed by enforcement authorities in respect of things done by them which they are required or authorised to do by or under this Act.

Regulations as to charges.

(2) Regulations under this section may include such provision as the Ministers see fit as regards charges for which the regulations provide and the recovery of such charges; and nothing in the following provisions shall prejudice this.

(3) Regulations under this section may provide that the amount of a charge (if imposed) is to be at the enforcement authority's discretion or to be at its discretion subject to a maximum or a minimum.

(4) Regulations under this section providing that a charge may not exceed a maximum amount, or be less than a minimum amount, may—

    (a) provide for one amount, or a scale of amounts to cover different prescribed cases; and

    (b) prescribe, as regards any amount, a sum or a method of calculating the amount.

46.—(1) Any expenses which are incurred under this Act by an authorised officer of a food authority in procuring samples, and causing samples to be analysed or examined, shall be defrayed by that authority.

Expenses of authorised officers and county councils.

(2) Any expenses incurred by a county council in the enforcement and execution of any provision of this Act, or of any regulations or orders made under it, shall, if the Secretary of State so directs, be defrayed as expenses for special county purposes charged on such part of the county as may be specified in the direction.

47. There shall be paid out of money provided by Parliament to the chairman of any tribunal constituted in accordance with regulations under this Act such remuneration (by way of salary or fees) and such allowances as the Ministers may with the approval of the Treasury determine.

Remuneration of tribunal chairmen.

Instruments and documents

48.—(1) Any power of the Ministers or the Minister to make regulations or an order under this Act includes power—

Regulations and orders.

    (a) to apply, with modifications and adaptations, any other enactment (including one contained in this Act) which deals with matters similar to those being dealt with by the regulations or order;

    (b) to make different provision in relation to different cases or classes of case (including different provision for different areas or different classes of business); and

    (c) to provide for such exceptions, limitations and conditions, and to make such supplementary, incidental, consequential or transitional provisions, as the Ministers or the Minister considers necessary or expedient.

(2) Any power of the Ministers or the Minister to make regulations or orders under this Act shall be exercisable by statutory instrument.

(3) Any statutory instrument containing—

(a) regulations under this Act; or

(b) an order under this Act other than an order under section 60(3) below,

shall be subject to annulment in pursuance of a resolution of either House of Parliament.

(4) Before making—

(a) any regulations under this Act, other than regulations under section 17(2) or 18(1)(c) above; or

(b) any order under Part I of this Act,

the Ministers shall consult with such organisations as appear to them to be representative of interests likely to be substantially affected by the regulations or order.

(5) Any consultation undertaken before the commencement of subsection (4) above shall be as effective, for the purposes of that subsection, as if undertaken after that commencement.

Form and authentication of documents.

49.—(1) The following shall be in writing, namely—

(a) all documents authorised or required by or under this Act to be given, made or issued by a food authority; and

(b) all notices and applications authorised or required by or under this Act to be given or made to, or to any officer of, such an authority.

(2) The Ministers may by regulations prescribe the form of any document to be used for any of the purposes of this Act and, if forms are so prescribed, those forms or forms to the like effect may be used in all cases to which those forms are applicable.

(3) Any document which a food authority are authorised or required by or under this Act to give, make or issue may be signed on behalf of the authority—

(a) by the proper officer of the authority as respects documents relating to matters within his province; or

(b) by any officer of the authority authorised by them in writing to sign documents of the particular kind or, as the case may be, the particular document.

(4) Any document purporting to bear the signature of an officer who is expressed—

(a) to hold an office by virtue of which he is under this section empowered to sign such a document; or

(b) to be duly authorised by the food authority to sign such a document or the particular document,

shall for the purposes of this Act, and of any regulations and orders made under it, be deemed, until the contrary is proved, to have been duly given, made or issued by authority of the food authority.

(5) In this section—

"proper officer", in relation to any purpose and to any food authority or any area, means the officer appointed for that purpose by that authority or, as the case may be, for that area;

"signature" includes a facsimile of a signature by whatever process reproduced.

50.—(1) Any document which is required or authorised by or under this Act to be given to or served on any person may, in any case for which no other provision is made by this Act, be given or served either—

(a) by delivering it to that person;

(b) in the case of any officer of an enforcement authority, by leaving it, or sending it in a prepaid letter addressed to him, at his office;

(c) in the case of an incorporated company or body, by delivering it to their secretary or clerk at their registered or principal office, or by sending it in a prepaid letter addressed to him at that office; or

(d) in the case of any other person, by leaving it, or sending it in a prepaid letter addressed to him, at his usual or last known residence.

(2) Where a document is to be given to or served on the owner or the occupier of any premises and it is not practicable after reasonable inquiry to ascertain the name and address of the person to or on whom it should be given or served, or the premises are unoccupied, the document may be given or served by addressing it to the person concerned by the description of "owner" or "occupier" of the premises (naming them) and—

(a) by delivering it to some person on the premises; or

(b) if there is no person on the premises to whom it can be delivered, by affixing it, or a copy of it, to some conspicuous part of the premises.

Amendments of other Acts

51.—(1) Part I of the Food and Environment Protection Act 1985 (contamination of food) shall have effect, and shall be deemed always to have had effect, subject to the amendments specified in subsection (2) below.

(2) The amendments referred to in subsection (1) above are—

(a) in subsection (1) of section 1 (power to make emergency orders), the substitution for paragraph (a) of the following paragraph—

"(a) there exist or may exist circumstances which are likely to create a hazard to human health through human consumption of food;";

(b) in subsection (2) of that section, the omission of the definition of "escape";

(c) the substitution for subsection (5) of that section of the following subsection—

"(5) An emergency order shall refer to the circumstances or suspected circumstances in consequence of which in the opinion of the designating authority making it food such as is mentioned in subsection (1)(b) above is, or may be, or may become, unsuitable for human consumption; and in this Act 'designated circumstances' means the circumstances or suspected circumstances to which an emergency order refers in pursuance of this subsection.";

(d) in section 2(3) (powers when emergency order has been made), the substitution for the words "a designated incident" of the words "designated circumstances";

(e) in paragraph (a) of subsection (1) of section 4 (powers of officers), the substitution for the words "an escape of substances" of the words "such circumstances as are mentioned in section 1(1) above"; and

(f) in paragraphs (b) and (c) of that subsection, the substitution for the words "the designated incident" of the words "the designated circumstances".

Markets, sugar beet and cold storage.
1984 c. 30.

52. In the Food Act 1984 (in this Act referred to as "the 1984 Act")—

(a) Part III (markets); and

(b) Part V (sugar beet and cold storage),

shall have effect subject to the amendments specified in Schedule 2 to this Act.

Supplemental

General interpretation.

53.—(1) In this Act, unless the context otherwise requires—

"the 1984 Act" means the Food Act 1984;

1956 c. 30.

"the 1956 Act" means the Food and Drugs (Scotland) Act 1956;

"advertisement" includes any notice, circular, label, wrapper, invoice or other document, and any public announcement made orally or by any means of producing or transmitting light or sound, and "advertise" shall be construed accordingly;

"analysis" includes microbiological assay and any technique for establishing the composition of food, and "analyse" shall be construed accordingly;

"animal" means any creature other than a bird or fish;

"article" does not include a live animal or bird, or a live fish which is not used for human consumption while it is alive;

"container" includes any basket, pail, tray, package or receptacle of any kind, whether open or closed;

"contravention", in relation to any provision, includes any failure to comply with that provision;

"cream" means that part of milk rich in fat which has been separated by skimming or otherwise;

"equipment" includes any apparatus;

1979 c. 2.

"exportation" and "importation" have the same meanings as they have for the purposes of the Customs and Excise Management Act 1979, and "export" and "import" shall be construed accordingly;

"fish" includes crustaceans and molluscs;

"functions" includes powers and duties;

"human consumption" includes use in the preparation of food for human consumption;

"knacker's yard" means any premises used in connection with the business of slaughtering, flaying or cutting up animals the flesh of which is not intended for human consumption;

"milk" includes cream and skimmed or separated milk;

"occupier", in relation to any ship or aircraft of a description specified in an order made under section 1(3) above or any vehicle, stall or place, means the master, commander or other person in charge of the ship, aircraft, vehicle, stall or place;

"officer" includes servant;

"preparation", in relation to food, includes manufacture and any form of processing or treatment, and "preparation for sale" includes packaging, and "prepare for sale" shall be construed accordingly;

"presentation", in relation to food, includes the shape, appearance and packaging of the food, the way in which the food is arranged when it is exposed for sale and the setting in which the food is displayed with a view to sale, but does not include any form of labelling or advertising, and "present" shall be construed accordingly;

"proprietor", in relation to a food business, means the person by whom that business is carried on;

"ship" includes any vessel, boat or craft, and a hovercraft within the meaning of the Hovercraft Act 1968, and "master" shall be construed accordingly; *1968 c. 59.*

"slaughterhouse" means a place for slaughtering animals, the flesh of which is intended for sale for human consumption, and includes any place available in connection with such a place for the confinement of animals while awaiting slaughter there or for keeping, or subjecting to any treatment or process, products of the slaughtering of animals there;

"substance" includes any natural or artificial substance or other matter, whether it is in solid or liquid form or in the form of a gas or vapour;

"treatment", in relation to any food, includes subjecting it to heat or cold.

(2) The following Table shows provisions defining or otherwise explaining expressions used in this Act (other than provisions defining or explaining an expression used only in the same section)—

authorised officer of a food authority	section 5(6)
business	section 1(3)
commercial operation	section 1(3) and (4)
contact material	section 1(3)
emergency control order	section 13(1)
emergency prohibition notice	section 12(1)
emergency prohibition order	section 12(2)
enforcement authority	section 6(1)
examination and examine	section 28(2)
food	section 1(1), (2) and (4)
food authority	section 5

(3) Any reference in this Act to regulations or orders made under it shall be construed as a reference to regulations or orders made under this Act by the Ministers or the Minister.

(4) For the purposes of this Act, any class or description may be framed by reference to any matters or circumstances whatever, including in particular, in the case of a description of food, the brand name under which it is commonly sold.

(5) Where, apart from this subsection, any period of less than seven days which is specified in this Act would include any day which is—

(a) a Saturday, a Sunday, Christmas Day or Good Friday; or

1971 c. 80.

(b) a day which is a bank holiday under the Banking and Financial Dealings Act 1971 in the part of Great Britain concerned,

that day shall be excluded from that period.

Application to Crown.

54.—(1) Subject to the provisions of this section, the provisions of this Act and of regulations and orders made under it shall bind the Crown.

(2) No contravention by the Crown of any provision of this Act or of any regulations or order made under it shall make the Crown criminally liable; but the High Court or, in Scotland, the Court of Session may, on the application of an enforcement authority, declare unlawful any act or omission of the Crown which constitutes such a contravention.

(3) Notwithstanding anything in subsection (2) above, the provisions of this Act and of regulations and orders made under it shall apply to persons in the public service of the Crown as they apply to other persons.

(4) If the Secretary of State certifies that it appears to him requisite or expedient in the interests of national security that the powers of entry conferred by section 32 above should not be exercisable in relation to any Crown premises specified in the certificate, those powers shall not be exercisable in relation to those premises; and in this subsection "Crown premises" means premises held or used by or on behalf of the Crown.

1947 c. 44.

(5) Nothing in this section shall be taken as in any way affecting Her Majesty in her private capacity; and this subsection shall be construed as if section 38(3) of the Crown Proceedings Act 1947 (interpretation of references in that Act to Her Majesty in her private capacity) were contained in this Act.

55.—(1) Nothing in Part II of this Act or any regulations or order made under that Part shall apply in relation to the supply of water to any premises, whether by a water undertaker or by means of a private supply (within the meaning of Chapter II of Part II of the Water Act 1989).

PART IV

Water supply: England and Wales.
1989 c. 15.

(2) In the following provisions of that Act, namely—

    section 52 (duties of water undertakers with respect to water quality);

    section 53 (regulations for preserving water quality); and

    section 64 (additional powers of entry for the purposes of Chapter II),

for the words "domestic purposes", wherever they occur, there shall be substituted the words "domestic or food production purposes".

(3) In subsection (2) of section 56 of that Act (general functions of local authorities in relation to water quality), for the words "domestic purposes" there shall be substituted the words "domestic or food production purposes" and for the words "those purposes" there shall be substituted the words "domestic purposes".

(4) In subsection (1) of section 57 of that Act (remedial powers of local authorities in relation to private supplies), for the words "domestic purposes", in the first place where they occur, there shall be substituted the words "domestic or food production purposes".

(5) In subsection (1) of section 66 of that Act (interpretation etc. of Chapter II), after the definition of "consumer" there shall be inserted the following definition—

    "'food production purposes' shall be construed in accordance with subsection (1A) below;".

(6) After that subsection there shall be inserted the following subsection—

    "(1A) In this Chapter references to food production purposes are references to the manufacturing, processing, preserving or marketing purposes with respect to food or drink for which water supplied to food production premises may be used; and in this subsection 'food production premises' means premises used for the purposes of a business of preparing food or drink for consumption otherwise than on the premises."

56.—(1) Nothing in Part II of this Act or any regulations or order made under that Part shall apply in relation to the supply of water to any premises, whether by a water authority (within the meaning of section 3 of the Water (Scotland) Act 1980) or by means of a private supply (within the meaning of Part VIA of that Act).

Water supply: Scotland.

1980 c. 45.

(2) In the following provisions of that Act, namely—

    section 76A (duties of water authorities with respect to water quality); and

    section 76B (regulations for preserving water quality),

for the words "domestic purposes", wherever they occur, there shall be substituted the words "domestic or food production purposes".

(3) In subsection (2) of section 76F of that Act (general functions of local authorities in relation to water quality), for the words "domestic purposes" there shall be substituted the words "domestic or food production purposes" and for the words "those purposes" there shall be substituted the words "domestic purposes".

(4) In subsection (1) of section 76G of that Act (remedial powers of local authorities in relation to private supplies), for the words "domestic purposes", in the first place where they occur, there shall be substituted the words "domestic or food production purposes".

(5) In subsection (1) of section 76L of that Act (interpretation etc. of Part VIA), after the definition of "analyse" there shall be inserted the following definition—

"'food production purposes' shall be construed in accordance with subsection (1A) below;".

(6) After that subsection there shall be inserted the following subsection—   .

"(1A) In this Part references to food production purposes are references to the manufacturing, processing, preserving or marketing purposes with respect to food or drink for which water supplied to food production premises may be used; and in this subsection 'food production premises' means premises used for the purposes of a business of preparing food or drink for consumption otherwise than on the premises."

Scilly Isles and Channel Islands.

57.—(1) This Act shall apply to the Isles of Scilly subject to such exceptions and modifications as the Ministers may by order direct.

(2) Her Majesty may by Order in Council direct that any of the provisions of this Act shall extend to any of the Channel Islands with such exceptions and modifications (if any) as may be specified in the Order.

Territorial waters and the continental shelf.

58.—(1) For the purposes of this Act the territorial waters of the United Kingdom adjacent to any part of Great Britain shall be treated as situated in that part.

1982 c. 23.

(2) An Order in Council under section 23 of the Oil and Gas (Enterprise) Act 1982 (application of civil law) may make provision for treating for the purposes of food safety legislation—

(a) any installation which is in waters to which that section applies; and

(b) any safety zone around any such installation,

as if they were situated in a specified part of the United Kingdom and for modifying such legislation in its application to such installations and safety zones.

(3) Such an Order in Council may also confer on persons of a specified description the right to require, for the purpose of facilitating the exercise of specified powers under food safety legislation—

(a) conveyance to and from any installation, including conveyance of any equipment required by them; and

(b) the provision of reasonable accommodation and means of subsistence while they are on any installation.

(4) In this section—

> "food safety legislation" means this Act and any regulations and orders made under it and any corresponding provisions in Northern Ireland;

> "installation" means an installation to which subsection (3) of the said section 23 applies;

> "safety zone" means an area which is a safety zone by virtue of Part III of the Petroleum Act 1987; and

> "specified" means specified in the Order in Council.

1987 c. 12.

59.—(1) The enactments mentioned in Schedule 3 to this Act shall have effect subject to the amendments there specified (being minor amendments and amendments consequential on the preceding provisions of this Act).

Amendments, transitional provisions, savings and repeals.

(2) The Ministers may by order make such modifications of local Acts, and of subordinate legislation (within the meaning of the Interpretation Act 1978), as appear to them to be necessary or expedient in consequence of the provisions of this Act.

1978 c. 30.

(3) The transitional provisions and savings contained in Schedule 4 to this Act shall have effect; but nothing in this subsection shall be taken as prejudicing the operation of sections 16 and 17 of the said Act of 1978 (which relate to the effect of repeals).

(4) The enactments mentioned in Schedule 5 to this Act (which include some that are spent or no longer of practical utility) are hereby repealed to the extent specified in the third column of that Schedule.

60.—(1) This Act may be cited as the Food Safety Act 1990.

Short title, commencement and extent.

(2) The following provisions shall come into force on the day on which this Act is passed, namely—

> section 13;

> section 51; and

> paragraphs 12 to 15 of Schedule 2 and, so far as relating to those paragraphs, section 52.

(3) Subject to subsection (2) above, this Act shall come into force on such day as the Ministers may by order appoint, and different days may be appointed for different provisions or for different purposes.

(4) An order under subsection (3) above may make such transitional adaptations of any of the following, namely—

> (a) the provisions of this Act then in force or brought into force by the order; and

> (b) the provisions repealed by this Act whose repeal is not then in force or so brought into force,

as appear to the Ministers to be necessary or expedient in consequence of the partial operation of this Act.

(5) This Act, except—

> this section;

> section 51,

section 58(2) to (4); and

paragraphs 7, 29 and 30 of Schedule 3 and, so far as relating to those paragraphs, section 59(1),

does not extend to Northern Ireland.

SCHEDULES

SCHEDULE 1

Section 16(3).

PROVISIONS OF REGULATIONS UNDER SECTION 16(1)

Composition of food

1. Provision for prohibiting or regulating—

    (a) the sale, possession for sale, or offer, exposure or advertisement for sale, of any specified substance, or of any substance of any specified class, with a view to its use in the preparation of food; or

    (b) the possession of any such substance for use in the preparation of food.

Fitness etc. of food

2.—(1) Provision for prohibiting—

    (a) the sale for human consumption; or

    (b) the use in the manufacture of products for sale for such consumption,

of food derived from a food source which is suffering or has suffered from, or which is liable to be suffering or to have suffered from, any disease specified in the regulations.

(2) Provision for prohibiting or regulating, or for enabling enforcement authorities to prohibit or regulate—

    (a) the sale for human consumption; or

    (b) the offer, exposure or distribution for sale for such consumption,

of shellfish taken from beds or other layings for the time being designated by or under the regulations.

3.—(1) Provision for regulating generally the treatment and disposal of any food—

    (a) which is unfit for human consumption; or

    (b) which, though not unfit for human consumption, is not intended for, or is prohibited from being sold for, such consumption.

(2) Provision for the following, namely—

    (a) for the registration by enforcement authorities of premises used or proposed to be used for the purpose of sterilising meat to which sub-paragraph (1) above applies, and for prohibiting the use for that purpose of any premises which are not registered in accordance with the regulations; or

    (b) for the issue by such authorities of licences in respect of the use of premises for the purpose of sterilising such meat, and for prohibiting the use for that purpose of any premises except in accordance with a licence issued under the regulations.

Processing and treatment of food

4. Provision for the following, namely—

    (a) for the giving by persons possessing such qualifications as may be prescribed by the regulations of written opinions with respect to the use of any process or treatment in the preparation of food, and for prohibiting the use for any such purpose of any process or treatment except in accordance with an opinion given under the regulations; or

Sᴄʜ. 1

(b) for the issue by enforcement authorities of licences in respect of the use of any process or treatment in the preparation of food, and for prohibiting the use for any such purpose of any process or treatment except in accordance with a licence issued under the regulations.

Food hygiene

5.—(1) Provision for imposing requirements as to—

(a) the construction, maintenance, cleanliness and use of food premises, including any parts of such premises in which equipment and utensils are cleaned, or in which refuse is disposed of or stored;

(b) the provision, maintenance and cleanliness of sanitary and washing facilities in connection with such premises; and

(c) the disposal of refuse from such premises.

(2) Provision for imposing requirements as to—

(a) the maintenance and cleanliness of equipment or utensils used for the purposes of a food business; and

(b) the use, for the cleaning of equipment used for milking, of cleaning agents approved by or under the regulations.

(3) Provision for requiring persons who are or intend to become involved in food businesses, whether as proprietors or employees or otherwise, to undergo such food hygiene training as may be specified in the regulations.

6.—(1) Provision for imposing responsibility for compliance with any requirements imposed by virtue of paragraph 5(1) above in respect of any premises—

(a) on the occupier of the premises; and

(b) in the case of requirements of a structural character, on any owner of the premises who either—

(i) lets them for use for a purpose to which the regulations apply; or

(ii) permits them to be so used after notice from the authority charged with the enforcement of the regulations.

(2) Provision for conferring in relation to particular premises, subject to such limitations and safeguards as may be specified, exemptions from the operation of specified provisions which—

(a) are contained in the regulations; and

(b) are made by virtue of paragraph 5(1) above,

while there is in force a certificate of the enforcement authority to the effect that compliance with those provisions cannot reasonably be required with respect to the premises or any activities carried on in them.

Inspection etc. of food sources

7.—(1) Provision for securing the inspection of food sources by authorised officers of enforcement authorities for the purpose of ascertaining whether they—

(a) fail to comply with the requirements of the regulations; or

(b) are such that any food derived from them is likely to fail to comply with those requirements.

(2) Provision for enabling such an officer, if it appears to him on such an inspection that any food source falls within sub-paragraph (1)(a) or (b) above, to give notice to the person in charge of the food source that, until a time specified in the notice or until the notice is withdrawn—

(a) no commercial operations are to be carried out with respect to the food source; and

(b) the food source either is not to be removed or is not to be removed except to some place so specified.

(3) Provision for enabling such an officer, if on further investigation it appears to him, in the case of any such food source which is a live animal or bird, that there is present in the animal or bird any substance whose presence is prohibited by the regulations, to cause the animal or bird to be slaughtered.

SCHEDULE 2

AMENDMENTS OF PARTS III AND V OF 1984 ACT

Amendments of Part III

1. Part III of the 1984 Act (markets) shall be amended in accordance with paragraphs 2 to 11 below.

2.—(1) In subsection (1) of section 50 (establishment or acquisition of markets), for the words "The council of a district" there shall be substituted the words "A local authority" and for the words "their district", in each place where they occur, there shall be substituted the words "their area".

(2) In subsection (2) of that section, for the words "the district" there shall be substituted the words "the authority's area".

(3) For subsection (3) of that section there shall be substituted the following subsection—

"(3) For the purposes of subsection (2), a local authority shall not be regarded as enjoying any rights, powers or privileges within another local authority's area by reason only of the fact that they maintain within their own area a market which has been established under paragraph (a) of subsection (1) or under the corresponding provision of any earlier enactment".

3. In section 51(2) (power to sell to local authority), the word "market" shall cease to have effect.

4.—(1) In subsection (1) of section 53 (charges by market authority), the words "and in respect of the weighing and measuring of articles and vehicles" shall cease to have effect.

(2) For subsection (2) of that section there shall be substituted the following subsection—

"(2) A market authority who provide—

(a) a weighing machine for weighing cattle, sheep or swine; or

(b) a cold air store or refrigerator for the storage and preservation of meat and other articles of food,

may demand in respect of the weighing of such animals or, as the case may be, the use of the store or refrigerator such charges as they may from time to time determine."

(3) In subsection (3)(b) of that section, the words "in respect of the weighing of vehicles, or, as the case may be," shall cease to have effect.

5. For subsection (2) of section 54 (time for payment of charges) there shall be substituted the following subsection—

"(2) Charges payable in respect of the weighing of cattle, sheep or swine shall be paid in advance to an authorised market officer by the person bringing the animals to be weighed."

SCH. 2 　　6. In section 56(1) (prohibited sales in market hours), for the word "district" there shall be substituted the word "area".

7. In section 57 (weighing machines and scales), subsection (1) shall cease to have effect.

8. After that section there shall be inserted the following section—

"Provision of
cold stores. 　　　57A.—(1) A market authority may provide a cold air store or refrigerator for the storage and preservation of meat and other articles of food.

(2) Any proposal by a market authority to provide under this section a cold air store or refrigerator within the area of another local authority requires the consent of that other authority, which shall not be unreasonably withheld.

(3) Any question whether or not such a consent is unreasonably withheld shall be referred to and determined by the Ministers.

1972 c. 70. 　　　(4) Subsections (1) to (5) of section 250 of the Local Government Act 1972 (which relate to local inquiries) shall apply for the purposes of this section as if any reference in those subsections to that Act included a reference to this section."

9. Section 58 (weighing of articles) shall cease to have effect.

10. In section 60 (market byelaws), after paragraph (c) there shall be inserted the following paragraph—

"(d) after consulting the fire authority for the area in which the market is situated, for preventing the spread of fires in the market."

11. In section 61 (interpretation of Part III), the words from "and this Part" to the end shall cease to have effect and for the definition of "market authority" there shall be substituted the following definitions—

1947 c. 41. 　　　"'fire authority' means an authority exercising the functions of a fire authority under the Fire Services Act 1947;

'food' has the same meaning as in the Food Safety Act 1990;

'local authority' means a district council, a London borough council or a parish or community council;

'market authority' means a local authority who maintain a market which has been established or acquired under section 50(1) or under the corresponding provisions of any earlier enactment."

Amendments of Part V

12. Part V of the 1984 Act (sugar beet and cold storage) shall be amended in accordance with paragraphs 13 to 16 below.

13.—(1) In subsections (1) and (2) of section 68 (research and education), for the word "Company", wherever it occurs, there shall be substituted the words "processors of home-grown beet".

(2) After subsection (5) of that section there shall be inserted the following subsection—

"(5A) An order under this section shall be made by statutory instrument which shall be subject to annulment in pursuance of a resolution of either House of Parliament.".

(3) In subsection (6) of that section, for the definition of "the Company" and subsequent definitions there shall be substituted—

"'year' means a period of 12 months beginning with 1st April;

and in this section and sections 69 and 69A 'home-grown beet' means sugar beet grown in Great Britain".

14. In subsection (3) of section 69 (crop price), for the words "'home-grown beet' means sugar beet grown in Great Britain; and" there shall be substituted the words "and section 69A".

15. After that section there shall be inserted the following section—

"Information.　　　　69A.—(1) For the purpose of facilitating—

(a) the making of a determination under section 69(1); or

(b) the preparation or conduct of discussions concerning Community arrangements for or relating to the regulation of the market for sugar,

the appropriate Minister may serve on any processor of home-grown beet a notice requiring him to furnish in writing, within such period as is specified in the notice, such information as is so specified.

(2) Subject to subsection (3), information obtained under subsection (1) shall not be disclosed without the previous consent in writing of the person by whom the information was furnished; and a person who discloses any information so obtained in contravention of this subsection shall be liable—

(a) on conviction on indictment, to a fine or to imprisonment for a term not exceeding two years or to both;

(b) on summary conviction, to a fine not exceeding the statutory maximum or to imprisonment for a term not exceeding three months or to both.

(3) Nothing in subsection (2) shall restrict the disclosure of information to any of the Ministers or the disclosure—

(a) of information obtained under subsection (1)(a)—

(i) to a person designated to make a determination under section 69(1); or

(ii) to a body which substantially represents the growers of home-grown beet; or

(b) of information obtained under subsection (1)(b), to the Community institution concerned.

(4) In this section "the appropriate Minister" means—

(a) in relation to England, the Minister of Agriculture, Fisheries and Food; and

(b) in relation to Scotland or Wales, the Secretary of State."

16. Section 70 (provision of cold storage) shall cease to have effect.

SCHEDULE 3

MINOR AND CONSEQUENTIAL AMENDMENTS

The Public Health Act 1936 (c. 49)

1. An order made by the Secretary of State under section 6 of the Public Health Act 1936 may constitute a united district for the purposes of any functions under this Act which are functions of a food authority in England and Wales.

The London Government Act 1963 (c. 33)

2. Section 54(1) of the London Government Act 1963 (food, drugs, markets and animals) shall cease to have effect.

The Agriculture Act 1967 (c. 22)

3. In section 7(3) of the Agriculture Act 1967 (labelling of meat in relation to systems of classifying meat), the words from "and, without prejudice" to the end shall cease to have effect.

4.—(1) In subsection (2) of section 25 of that Act (interpretation of Part I), for the definition of "slaughterhouse" there shall be substituted the following definition—

1974 c. 3.
1980 c. 13.

"'slaughterhouse' has, in England and Wales, the meaning given by section 34 of the Slaughterhouses Act 1974 and, in Scotland, the meaning given by section 22 of the Slaughter of Animals (Scotland) Act 1980;".

(2) In subsection (3) of that section, for the words from "Part II" to "1955" there shall be substituted the words "section 15 of the Slaughterhouses Act 1974 or section 1 of the Slaughter of Animals (Scotland) Act 1980".

The Farm and Garden Chemicals Act 1967 (c. 50)

5. In section 4 of the Farm and Garden Chemicals Act 1967 (evidence of analysis of products)—

1984 c. 30.

(a) in subsection (3), for the words "section 76 of the Food Act 1984" there shall be substituted the words "section 27 of the Food Safety Act 1990"; and

(b) in subsection (7)(c), the words from "for the reference" to "1956" shall cease to have effect.

The Trade Descriptions Act 1968 (c. 29)

6. In section 2(5)(a) of the Trade Descriptions Act 1968 (certain descriptions to be deemed not to be trade descriptions), for the words "the Food Act 1984, the Food and Drugs (Scotland) Act 1956" there shall be substituted the words "the Food Safety Act 1990".

1956 c. 30.

7. In section 22 of that Act (admissibility of evidence in proceedings for offences under Act), in subsection (2), the paragraph beginning with the words "In this subsection" shall cease to have effect, and after that subsection there shall be inserted the following subsection—

"(2A) In subsection (2) of this section—

1968 c. 67.
S.I. 1989/846
(N.I.6).

'the food and drugs laws' means the Food Safety Act 1990, the Medicines Act 1968 and the Food (Northern Ireland) Order 1989 and any instrument made thereunder;

'the relevant provisions' means—

(i) in relation to the said Act of 1990, section 31 and regulations made thereunder;

(ii) in relation to the said Act of 1968, so much of Schedule 3 to that Act as is applicable to the circumstances in which the sample was procured; and

(iii) in relation to the said Order, Articles 40 and 44,

or any provisions replacing any of those provisions by virtue of section 17 of the said Act of 1990, paragraph 27 of Schedule 3 to the said Act of 1968 or Article 72 or 73 of the said Order."

The Medicines Act 1968 (c. 67)

8. In section 108 of the Medicines Act 1968 (enforcement in England and Wales)—

> (a) for the words "food and drugs authority", in each place where they occur, there shall be substituted the words "drugs authority"; and

> (b) after subsection (11) there shall be inserted the following subsection—

>> "(12) In this section 'drugs authority' means—

>>> (a) as respects each London borough, metropolitan district or non-metropolitan county, the council of that borough, district or county; and

>>> (b) as respects the City of London (including the Temples), the Common Council of that City."

9. In section 109 of that Act (enforcement in Scotland)—

> (a) paragraph (c) of subsection (2) shall cease to have effect; and

> (b) after that subsection there shall be inserted the following subsection—

>> "(2A) Subsection (12) of section 108 of this Act shall have effect in relation to Scotland as if for paragraphs (a) and (b) there were substituted the words "an islands or district council".

10. After section 115 of that Act there shall be inserted the following section—

"Facilities for microbiological examinations.
115A. A drugs authority cr the council of a non-metropolitan district may provide facilities for microbiological examinations of drugs."

11. In section 132(1) of that Act (interpretation), the definition of "food and drugs authority" shall cease to have effect and after the definition of "doctor" there shall be inserted the following definition—

> "'drugs authority' has the meaning assigned to it by section 108(12) of this Act;".

12. In paragraph 1(2) of Schedule 3 to that Act (sampling) for the words from "in relation to England and Wales" to "Food and Drugs (Scotland) Act 1956" there shall be substituted the words "except in relation to Northern Ireland, has the meaning assigned to it by section 27 of the Food Safety Act 1990".

The Transport Act 1968 (c. 73)

13. In Schedule 16 to the Transport Act 1968 (supplementary and consequential provisions), in paragraph 7(2), paragraphs (d) and (e) shall cease to have effect.

The Tribunals and Inquiries Act 1971 (c. 62)

14.—(1) In Schedule 1 to the Tribunals and Inquiries Act 1971 (tribunals under supervision of Council on Tribunals), paragraph 15 shall cease to have effect and after paragraph 6B there shall be inserted the following paragraph—

| "Food | 6C. Tribunals constituted in accordance with regulations under Part II of the Food Safety Act 1990." |

SCH. 3   (2) In that Schedule, paragraph 40 shall cease to have effect and after paragraph 36 there shall be inserted the following paragraph—

| "Food | 36A. Tribunals constituted in accordance with regulations under Part II of the Food Safety Act 1990 being tribunals appointed for Scotland." |

The Agriculture (Miscellaneous Provisions) Act 1972 (c. 62)

15.—(1) In subsection (1) of section 4 of the Agriculture (Miscellaneous Provisions) Act 1972 (furnishing by milk marketing boards of information derived from tests of milk)—

(a) for the words "appropriate authority" there shall be substituted the words "enforcement authority"; and

(b) for the words from "Milk and Dairies Regulations" to "1956" there shall be substituted the words "regulations relating to milk, dairies or dairy farms which were made under, or have effect as if made under, section 16 of the Food Safety Act 1990."

(2) In subsection (2) of that section, for the definition of "appropriate authority" there shall be substituted the following definition—

"'enforcement authority' has the same meaning as in the Food Safety Act 1990;".

(3) Subsection (3) of that section shall cease to have effect.

The Poisons Act 1972 (c. 66)

1956 c. 30.   16. In section 8(4)(a) of the Poisons Act 1972 (evidence of analysis in proceedings under Act) for the words "section 76 of the Food Act 1984, or section 27 of the Food and Drugs (Scotland) Act 1956" there shall be substituted the words "section 27 of the Food Safety Act 1990".

The Local Government Act 1972 (c. 70)

17. In section 259(3) of the Local Government Act 1972 (compensation for loss of office)—

(a) in paragraph (b), for the words "food and drugs authority, within the meaning of the Food Act 1984" there shall be substituted the words "food authority within the meaning of the Food Safety Act 1990";

(b) in paragraph (c), for sub-paragraphs (i) and (ii) there shall be substituted the words "which are incorporated or reproduced in the Slaughterhouses Act 1974 or the Food Safety Act 1990"; and

1955 c. 16.   (c) the words "section 129(1) of the Food and Drugs Act 1955" shall cease to have effect.

The Slaughterhouses Act 1974 (c. 3)

18. In the following provisions of the Slaughterhouses Act 1974, namely—

(a) section 2(2)(a) (requirements to be complied with in relation to slaughterhouse licences);

(b) section 4(2)(a) (requirements to be complied with in relation to knacker's yard licences);

(c) section 12(2) (regulations with respect to slaughterhouses and knackers' yards to prevail over byelaws); and

(d) section 16(3) (regulations with respect to public slaughterhouses to prevail over byelaws),

for the words "section 13 of the Food Act 1984" there shall be substituted the words "section 16 of the Food Safety Act 1990".

The Licensing (Scotland) Act 1976 (c. 66)

19. In section 23(4) of the Licensing (Scotland) Act 1976 (application for new licence), for the words "section 13 of the Food and Drugs (Scotland) Act 1956" there shall be substituted "section 16 of the Food Safety Act 1990".

The Weights and Measures &c. Act 1976 (c. 77)

20.—(1) In subsection (1) of section 12 of the Weights and Measures &c. Act 1976 (shortages of food and other goods), for paragraphs (a) and (b) there shall be substituted the following paragraph—

"(a) section 16 of the Food Safety Act 1990 ('the 1990 Act');".

(2) In subsection (9) of that section—

(a) for paragraph (a) there shall be substituted the following paragraph—

"(a) where it was imposed under the 1990 Act—

(i) the Minister of Agriculture, Fisheries and Food and the Secretary of State acting jointly in so far as it was imposed in relation to England and Wales; and

(ii) the Secretary of State in so far as it was imposed in relation to Scotland;"; and

(b) in paragraph (c), the words "the 1956 Act or" shall cease to have effect.

21. In Schedule 6 to that Act (temporary requirements imposed by emergency orders), for paragraphs 2 and 3 there shall be substituted the following paragraph—

"Food Safety Act 1990 (c. 16)

2.—(1) This paragraph applies where the relevant requirement took effect under or by virtue of the Food Safety Act 1990.

(2) The following provisions of that Act—

(a) Part I (preliminary);

(b) Part III (administration and enforcement); and

(c) sections 40 to 50 (default powers and other supplemental provisions),

shall apply as if the substituted requirement were imposed by regulations under section 16 of that Act."

The Hydrocarbon Oil Duties Act 1979 (c. 5)

22. In Schedule 5 to the Hydrocarbon Oil Duties Act 1979 (sampling) in paragraph 5(d) for the words "section 76 of the Food Act 1984, section 27 of the Food and Drugs (Scotland) Act 1956" there shall be substituted the words "section 27 of the Food Safety Act 1990".

The Slaughter of Animals (Scotland) Act 1980 (c. 13)

23. In section 19(2) of the Slaughter of Animals (Scotland) Act 1980 (enforcement) for the words "section 13 of the Food and Drugs (Scotland) Act 1956" there shall be substituted the words "section 16 of the Food Safety Act 1990" and for the words "section 36 of the said Act of 1956" there shall be substituted the words "section 32 of the said Act of 1990".

SCH. 3   24. In section 22 of that Act (interpretation)—

(a) for the definition of "knacker's yard" there shall be substituted the following definition—

"'knacker's yard' means any premises used in connection with the business of slaughtering, flaying or cutting up animals the flesh of which is not intended for human consumption; and 'knacker' means a person whose business it is to carry out such slaughtering, flaying or cutting up"; and

(b) for the definition of "slaughterhouse" there shall be substituted the following definition—

"'slaughterhouse' means a place for slaughtering animals, the flesh of which is intended for human consumption, and includes any place available in connection with such a place for the confinement of animals while awaiting slaughter there or keeping, or subjecting to any treatment or process, products of the slaughtering of animals there; and 'slaughterman' means a person whose business it is to carry out such slaughtering".

The Civic Government (Scotland) Act 1982 (c. 45)

25. In section 39 of the Civic Government (Scotland) Act 1982 (street traders' licences)—

1914 c. 46.

(a) in subsection (3)(b), for the words "section 7 of the Milk and Dairies (Scotland) Act 1914" there shall be substituted the words "regulations made under section 19 of the Food Safety Act 1990"; and

(b) in subsection (4)—

1956 c. 30.

(i) for the words "regulations made under sections 13 and 56 of the Food and Drugs (Scotland) Act 1956", there shall be substituted the words "section 1(3) of the Food Safety Act 1990";

(ii) for the words "islands or district council" there shall be substituted the words "food authority (for the purposes of section 5 of the Food Safety Act 1990)"; and

(iii) for the words "sections 13 and 56 of the Food and Drugs (Scotland) Act 1956", there shall be substituted the words "section 16 of the Food Safety Act 1990".

The Public Health (Control of Disease) Act 1984 (c. 22)

26. In section 3(2) of the Public Health (Control of Disease) Act 1984 (jurisdiction and powers of port health authority), for paragraph (a) there shall be substituted the following paragraph—

"(a) of a food authority under the Food Safety Act 1990;".

27. In section 7(3) of that Act (London port health authority), for paragraph (d) there shall be substituted the following paragraph—

"(d) of a food authority under any provision of the Food Safety Act 1990."

28.—(1) In subsection (1) of section 20 of that Act (stopping of work to prevent spread of disease), in paragraph (b) for the words "subsection (1) of section 28 of the Food Act 1984" there shall be substituted "subsection (1A) below".

1984 c. 30.

(2) After that subsection there shall be inserted the following subsection—

"(1A) The diseases to which this subsection applies are—

(a) enteric fever (including typhoid and paratyphoid fevers);

(b) dysentery;

(c) diphtheria;

(d) scarlet fever;

(e) acute inflammation of the throat;

(f) gastro-enteritis; and

(g) undulant fever."

The Food and Environment Protection Act 1985 (c. 48)

29. In section 24(1) of the Food and Environment Protection Act 1985 (interpretation)—

(a) in the definition of "designated incident", for the words "designated incident" there shall be substituted the words "designated circumstances";

(b) the definition of "escape" shall cease to have effect; and

(c) for the definition of "food" there shall be substituted—

"'food' has the same meaning as in the Food Safety Act 1990."

30. In section 25 of that Act (Northern Ireland) after subsection (4) there shall be inserted the following subsection—

"(4A) Section 24(1) above shall have effect in relation to Northern Ireland as if for the definition of 'food' there were substituted the following definition—

'"food" has the meaning assigned to it by Article 2(2) of the Food (Northern Ireland) Order 1989, except that it includes water which is bottled or is an ingredient of food;'."

S.I. 1989/846 (N.I.6).

The Local Government Act 1985 (c. 51)

31. In paragraph 15 of Schedule 8 to the Local Government Act 1985 (trading standards and related functions)—

(a) sub-paragraph (2) shall cease to have effect; and

(b) at the end of sub-paragraph (6) there shall be added the words "or section 5(1) of the Food Safety Act 1990".

The Weights and Measures Act 1985 (c. 72)

32. In section 38 of the Weights and Measures Act 1985 (special powers of inspectors), subsection (4) (exclusion for milk) shall cease to have effect.

33. In section 93 of that Act (powers under other Acts with respect to marking of food) for the words "Food Act 1984" there shall be substituted the words "Food Safety Act 1990".

1984 c. 30.

34. In section 94(1) of that Act (interpretation), in the definition of "drugs" and "food" for the words "Food Act 1984, or, in Scotland, the Food and Drugs (Scotland) Act 1956" there shall be substituted the words "Food Safety Act 1990".

1956 c. 30.

The Agriculture Act 1986 (c. 49)

35. In section 1(6) of the Agriculture Act 1986 (provision of agricultural goods and services), in the definition of "food", for the words "Food Act 1984" there shall be substituted "Food Safety Act 1990".

The National Health Service (Amendment) Act 1986 (c. 66)

36.—(1) In subsection (2) of section 1 of the National Health Service (Amendment) Act 1986 (application of food legislation to health authorities and health service premises)—

(a) for the words "appropriate authority" there shall be substituted the word "Ministers"; and

(b) for the word "authority" there shall be substituted the word "Ministers".

(2) For subsection (7) of that section there shall be substituted—

"(7) In this section—

'the Ministers' has the same meaning as in the Food Safety Act 1990;

'the food legislation' means the Food Safety Act 1990 and any regulations or orders made (or having effect as if made) under it;

'health authority'—

(a) as respects England and Wales, has the meaning assigned to it by section 128 of the 1977 Act; and

1984 c. 36.

(b) as respects Scotland, means a Health Board constituted under section 2 of the 1978 Act, the Common Services Agency constituted under section 10 of that Act or a State Hospital Management Committee constituted under section 91 of the Mental Health (Scotland) Act 1984."

The Consumer Protection Act 1987 (c. 43)

37. In section 19(1) of the Consumer Protection Act 1987 (interpretation of Part II), in the definition of "food" for the words "Food Act 1984" there shall be substituted "Food Safety Act 1990".

The Road Traffic Offenders Act 1988 (c. 53)

1956 c. 30.

38. In section 16(7) of the Road Traffic Offenders Act 1988 (meaning of "authorised analyst" in relation to proceedings under Act), for the words "section 76 of the Food Act 1984, or section 27 of the Food and Drugs (Scotland) Act 1956" there shall be substituted the words "section 27 of the Food Safety Act 1990".

Section 59(3).

SCHEDULE 4

Transitional Provisions and Savings

Ships and aircraft

1. In relation to any time before the commencement of the first order under section 1(3) of this Act—

(a) any ship which is a home-going ship within the meaning of section 132 of the 1984 Act or section 58 of the 1956 Act (interpretation) shall be regarded as premises for the purposes of this Act; and

(b) the powers of entry conferred by section 32 of this Act shall include the right to enter any ship or aircraft for the purpose of ascertaining whether there is in the ship or aircraft any food imported as part of the cargo in contravention of the provisions of regulations made under Part II of this Act;

and in this Act as it applies by virtue of this paragraph "occupier", in relation to any ship or aircraft, means the master, commander or other person in charge of the ship or aircraft.

Regulations under the 1984 Act

2.—(1) In so far as any existing regulations made, or having effect as if made, under any provision of the 1984 Act specified in the first column of Table A below have effect in relation to England and Wales, they shall have effect, after the commencement of the relevant repeal, as if made under the provisions of this Act specified in relation to that provision in the second column of that Table, or such of those provisions as are applicable.

(2) In this paragraph and paragraphs 3 and 4 below "existing regulations" means—

   (a) any regulations made, or having effect as if made, under a provision repealed by this Act; and

   (b) any orders having effect as if made under such regulations,

which are in force immediately before the coming into force of that repeal; and references to the commencement of the relevant repeal shall be construed accordingly.

TABLE A

Provision of the 1984 Act	Provision of this Act
section 4 (composition etc. of food)	sections 16(1)(a), (c) and (f) and (3) and 17(1)
section 7 (describing food)	section 16(1)(e)
section 13 (food hygiene)	section 16(1)(b), (c), (d) and (f), (2) and (3)
section 33 (milk and dairies)	section 16(1)(b), (c), (d) and (f), (2) and (3)
section 34 (registration), so far as relating to dairies or dairy farms	section 19
section 38 (milk: special designations)	section 18(2)
section 73(2) (qualification of officers)	section 5(6)
section 76(2) (public analysts)	section 27(2)
section 79(5) (form of certificate)	section 49(2)
section 119 (Community provisions)	section 17(2)

Regulations under the 1956 Act

3. Any existing regulations made, or having effect as if made, under any provision of the 1956 Act specified in the first column of Table B below shall have effect, after the commencement of the relevant repeal, as if made under the provisions of this Act specified in relation to that provision in the second column of that Table, or such of those provisions as are applicable.

SCH. 4   TABLE B

Provision of the 1956 Act	Provision of this Act
section 4 (composition etc. of food)	sections 16(1)(a), (c) and (f) and (3) and 17(1)
section 7 (describing food)	section 16(1)(e)
section 13 (food hygiene)	sections 5(6) and 16(1)(b), (c), (d) and (f), (2) and (3)
section 16(2) (regulations as to milk)	section 18(2)
section 27(2) (public analysts)	section 27(2)
section 29(3) (form of certificate)	section 49(2)
section 56A (Community provisions)	section 17(2)

Other regulations

1983 c. 37.

4. In so far as any existing regulations made under section 1 of the Importation of Milk Act 1983 have effect in relation to Great Britain, they shall have effect, after the commencement of the relevant repeal, as if made under section 18(1)(c) of this Act.

Orders with respect to milk in Scotland

1914 c. 46.

5.—(1) Any existing order made under section 12(2) of the Milk and Dairies (Scotland) Act 1914 (orders with respect to milk) shall have effect, after the commencement of the relevant repeal, as if it were regulations made under section 16(1)(b), (d) and (f) and (2) of this Act.

1922 c. 54.

(2) Any existing order made under section 3 of the Milk and Dairies (Amendment) Act 1922 (sale of milk under special designations) shall have effect, after the commencement of the relevant repeal, as if it were regulations made under section 18(2) of this Act.

(3) In this paragraph "existing order" means any order made under a provision repealed by this Act which is in force immediately before the coming into force of that repeal; and references to the commencement of the relevant repeal shall be construed accordingly.

Disqualification orders

6. The repeal by this Act of section 14 of the 1984 Act (court's power to disqualify caterers) shall not have effect as respects any order made, or having effect as if made, under that section which is in force immediately before the commencement of that repeal.

Food hygiene byelaws

7.—(1) The repeal by this Act of section 15 of the 1984 Act (byelaws as to food) shall not have effect as respects any byelaws made, or having effect as if made, under that section which are in force immediately before the commencement of that repeal.

(2) In so far as any such byelaws conflict with any regulations made, or having effect as if made, under Part II of this Act, the regulations shall prevail.

Closure orders

8. The repeal by this Act of section 21 of the 1984 Act or section 1 of the Control of Food Premises (Scotland) Act 1977 (closure orders) shall not have effect as respects any order made, or having effect as if made, under that section which is in force immediately before the commencement of that repeal.

1977 c. 28.

SCHEDULE 5

REPEALS

Chapter	Short title	Extent of repeal
1914 c. 46.	The Milk and Dairies (Scotland) Act 1914.	The whole Act.
1922 c. 54.	The Milk and Dairies (Amendment) Act 1922.	The whole Act.
1934 c. 51.	The Milk Act 1934.	The whole Act.
1949 c. 34.	The Milk (Special Designations) Act 1949.	The whole Act.
1956 c. 30.	The Food and Drugs (Scotland) Act 1956.	The whole Act.
1963 c. 33.	The London Government Act 1963.	Section 54(1).
1967 c. 22.	The Agriculture Act 1967.	In section 7(3), the words from "and, without prejudice" to the end.
1967 c. 50.	The Farm and Garden Chemicals Act 1967.	In section 4(7)(c), the words from "for the reference" to "1956".
1968 c. 29.	The Trade Descriptions Act 1968.	In section 22(2), the paragraph beginning with the words "In this subsection".
1968 c. 67.	The Medicines Act 1968.	In section 132(1), the definition of "food and drugs authority". In Schedule 5, paragraph 17.
1968 c. 73.	The Transport Act 1968.	In Schedule 16, in paragraph 7(2), paragraphs (d) and (e).
1971 c. 62.	The Tribunals and Inquiries Act 1971.	In Schedule 1, paragraphs 15 and 40.
1972 c. 66.	The Agriculture (Miscellaneous Provisions) Act 1972.	Section 4(3).
1972 c. 68.	The European Communities Act 1972.	In Schedule 4, paragraph 3(2)(c).
1976 c. 77.	The Weights and Measures &c. Act 1976.	In section 12(9)(c), the words "the 1956 Act or".
1977 c. 28.	The Control of Food Premises (Scotland) Act 1977.	The whole Act.
1983 c. 37.	The Importation of Milk Act 1983.	The whole Act.
1984 c. 30.	The Food Act 1984.	Parts I and II. In section 51(2), the word "market". In section 53, in subsection (1) the words "and in respect of the weighing

Chapter	Short title	Extent of repeal
		and measuring of articles and vehicles", and in subsection (3)(b) the words "in respect of the weighing of vehicles, or as the case may be," Section 57(1). Section 58. In section 61, the words from "and this Part" to the end. Part IV. Sections 70 to 92. In section 93, in subsection (2), paragraphs (b) to (d) and, in subsection (3), paragraphs (a) to (e) and (h) to (l). In section 94, subsection (1) except as regards offences under Part III of the Act, and subsection (2). In section 95, subsections (2) to (8). Sections 96 to 109. Sections 111 to 120. In section 121, subsections (2) and (3). Sections 122 to 131. In section 132, subsection (1) except the words "In this Act, unless the context otherwise requires" and the definitions of "animal" and "the Minister". Sections 133 and 134. In section 136, in subsection (2), paragraphs (b) and (c). Schedules 1 to 11.
1985 c. 48.	The Food and Environment Protection Act 1985.	In section 1(2), the definition of "escape". In section 24(1), the definition of "escape".
1985 c. 51.	The Local Government Act 1985.	In Schedule 8, paragraph 15(2).
1985 c. 72.	The Weights and Measures Act 1985.	Section 38(4).

PRINTED IN THE UNITED KINGDOM BY MIKE LYNN
Controller and Chief Executive of Her Majesty's Stationery Office
and Queen's Printer of Acts of Parliament.
1st Impression July 1990
6th Impression December 1995

Printed in the United Kingdom for HMSO
Dd 5064346 C15 12/95 1731 56219 ON 322071